Interview with a Wizard
Peter J Carroll

Interview with
Ian Blumberg-Enge

Copyright © 2022 Peter J Carroll

First edition

All rights reserved. No part of this work may be reproduced or utilized in any form by any means electronic or mechanical, including *xerography, photocopying, microfilm,* and *recording,* or by any information storage system without permission in writing from the publishers.

Other books by Peter J Carroll and published by Mandrake
Octavo: A Sorcerer-Scientist's Grimoire
Apophenian: A Chaos Magic Paradigm

Contents

Introduction 7
Foreword by Lionel Snell 8
Preface by Ian Blumberg-Enge 11
Preface by Peter J Carroll 12

(For list of questions & topics covered see page 303.
below is one randon example from each interview)

Interview 1 13
1) Cosmology, The Magical origins of Hypersphere Cosmology.

Interview 2 31
38) Yoruba Practices.

Interview 3 62
102) Twin Towers conspiracy?

Interview 4 77
133) Magical effects.

Interview 5 105
165) Warp Drive.

Interview 6 125
361) Optimist or Pessimist?

Interview 7 146
184) The Magical Diary.

Interview 8 164
226) Universal Basic Income.

Interview 9 185
253) Meditation and Compassion.

Interview 10 205
285) National Exorcism & Humour.

Interview 11 224
305) Poetry & Slang.

Interview 12 244
327) Sexism, Religion, and Civilisation.

Interview 13 263
348) Simulation Theory?

Appendix, The Occultaris part 1 272
Key to algebraic symbols 292
Formulae 293
The CMBR 299
Index of Questions 304
Index 318

Introduction

Peter J Carroll developed the Theories and Practices of Chaos Magic which caused a revolution in magical and esoteric thinking in the last few decades of the twentieth century, and they continue to heavily influence it in the twenty first century.

As the first in a new tradition of Sorcerer-Scientists, Carroll developed a paradigm in which Magic lies far closer to Science than to Religion. He formed a worldwide magical order to promote and develop the new insights into magic and then largely disappeared from public view to procreate and to build a business empire and continue scientific and metaphysical research in private.

This book of interviews reveals the extraordinary life and adventures and thoughts of one of the world's most exceptional antinomian thinkers. It contains much that may intrigue, surprise, amaze, and outrage the reader on a very wide range of topics.

The appendix contains Carroll's Hypersphere Cosmology thesis which presents a radical alternative to the big-bang theory, and which has, so far, resisted all attempts to falsify it.

Foreword by Lionel Snell

This book is an extended interview with Peter James Carroll, magician, author, founder of the Illuminates of Thanateros, pioneering co-creator of chaos magic practice and theory and – as we also find out – rebel physicist.

It is one hell of an interview, comprising in-depth responses to no less than 365 questions. I'd say the interviewer has not simply consulted his subject but effectively taken him apart, inspected the parts and reassembled them to reveal a rich and comprehensive picture of not just an original thinker, but also an effective change-maker and a human being.

The topics range from Magic to Science and Religion, Travel, Ritual, Conspiracy, Sex Magic, LSD, Parenting… It is simply ridiculous of me to attempt a list. Just turn to the Contents List on page XX and browse the topics.

Pete is above all respected as a "Results" magician. Magic only just survived the highly sceptical 1950s by adopting a mask of psychology – as Dion Fortune adapted Crowley's definition: from the Science and Art of "causing Change" to occur in conformity with Will to causing "changes in consciousness". In my 1970s book *SSOTBME An Essay on Magic* I paved the way to Results magic by suggesting just how far a change in consciousness can actually lead to outward change. That was why Gerald Suster described it as "the book that put the magic back in magic". But it was Pete and his fellow Chaos pioneers that really brought Results Magic to the fore.

There is plenty in this book for those whose main interest is in Results Magic – practical advice, suggested techniques and accounts of successful operations. But I also find much more: the book gives a real feel for what it means to think, live and *experience magic* – as well as to *do magic*.

I have never been one for attempting to define magic, but I do enjoy describing it from varied viewpoints. One such is to see magic as the path or journey between Science and Art. At the Science end, where I most clearly see Crowley and Pete, magic is something you *do to cause change*. Towards the Art end it becomes a way to *live and experience change*.

It is the difference between: A) the person who observes waves crashing against the shore and seeks ways to resist their impact, or ways to harvest their energy or, at its most absurd, to sit like King Canute and try to order the sea to retreat; and B) the person who risks surfing the waves to revel in their sheer chaos as an experience of learning and growing. Living magically is rich, and sometimes unsettling.

If Results magic is the Way of Power, then Living magically could be the Way of Strength – and there is plenty of both in this volume.

There is also a major section that is for me the best yet introduction to Pete's Hypersphere Cosmology. Like others, I have struggled with this in the past as I tried to get a handle on rotation about a plane in 4-space. I still struggle – but now I wish to know more.

It presents an interesting example of a relationship between Science and Magic. Science, as a child of the Protestant Enlightenment, has no time for heretics. Where Pete presents powerful arguments, their very power can be seen as a diabolical delusion. His theory will be rejected as "magic" until such time as physical experimentation suggests that the hypersphere model is more "true" than current beliefs. Then the theory will suddenly be transmuted from "magic" into "science" – and the experimenters will claim all the credit. Scientists sees their world as a pool of bright light, while magicians explore the shadows beyond.

To find out more – much more – read on. You could dip into the

Contents list and choose topics of interest, or read it cover to cover as I did. With 365 questions responded to, you could even spend an initiatory year reading one question each day and meditating on its implications. Do what thou wilt!

Preface by Ian Blumberg-Enge

In late 2020, like many people, I was rekindling my occult studies to fill the free alone time of the global covid-19 pandemic. At that point I'd been studying magic in general and chaos magic in particular on and off for about 13 years. while I consider myself one of those people who has too much trouble with everyday life to actually practice magic, the implications for psychology, philosophy and physics make it an undeniably important subject and so I learned the theory and history fairly well.

One night while googling around I stumbled across Pete's website. I had read Liber Null at a young age and while it's a revolutionary work it felt very steeped in the western occult tradition and so I was surprised to find that Pete's website was not only not particularly magic focused but his latest project was a cosmological model! I immediately had a million questions and to my delight Pete quickly agreed to an interview.

Initially I figured we'd do a few thousand words, whatever was average for a periodical, but after that was quickly finished and I had even more questions I pitched Pete the idea of trying to fill a full book and to my continued amazement he agreed to that too. The resulting conversation challenged a lot of my preconceived notions about chaos magic, Peter Carroll, and the western occult tradition in the most interesting, fun, and often funny way possible. I will be forever grateful to Pete for the opportunity and experience and I hope readers will enjoy it as much as I have.

Preface by Peter J Carroll

Ian Blumberg contacted me advertising himself as a budding journalist of the esoteric and asked for an e-mail interview. WTFN? I asked myself, and we both got carried away with the project which reached book length after many months of daily exchanges. Herewith the result, which meanders all over the place from biography to magical theory, to politics, religion, quantum science, sex magic, economics, sociology, magical practise, travelogue, and adventure, to business, education, ufology, philosophy, and cosmology. Perhaps we have between us invented the literary equivalent of a late night chat show series. I would recommend reading one interview per evening.

Interview 1

1) Maybe we could start with your current project, an alternative theory to the big bang? Is it possible to give a basic description? What is wrong with or missing from the big bang? How did you get into this project?

Back in the 1980s and 90s a flurry of popular science books came out on the Big Bang theory. At the time I had been conducting a series of seminars and tours about Chaos Magic and founding an international chaos magic order – The IOT. Several participants suggested parallels between the idea of a big bang explosive beginning to the universe and the idea of a primal chaotic creativity underling the mind and the universe, and the ancient Greek idea of the universe arising from Chaos.

At an Austrian Castle we conducted a mass dreamwork scrying experiment following a ritual to send an entity back to the time of the big bang to have a look at it.

Strangely, most participants reported that the universe then looked broadly the same as it does now.

This sent me on a twenty-year quest to resolve the issue. I had a science degree but in Chemistry rather than Physics. Nevertheless, I plunged into the study of cosmology to discover in detail just why so many physicists supported the Big Bang theory. This quest took me through thousands of scientific papers and books and entailed a substantial upgrade to my maths skills.

What I found rather shocked me and goaded me to further efforts. The Big Bang theory had become concocted in defiance of scientific method; it began with an experimental assumption about the redshift of distant galaxies and all subsequent observations have become interpreted in terms of that assumption, no matter how awkward the fit. Later observations that seemed to contradict the initial assumption did not provoke a significant questioning of it, instead they became explained away by inventing vast imaginary cosmic phenomena like so-called dark matter and dark energy. In any other field of science, you could not get away with this, much less get Nobel Prizes for it.

I found it increasingly irritating to find so many cosmologists and popularisers asserting that 'We KNOW that the universe is expanding'. I also found the Big Bang theory increasingly unsatisfactory in metaphysical terms for it implied a creation event and an eventual apocalypse which gave some comfort to those with a monotheist religious outlook. (One of the main initial architects of the theory - Georges Lemaitre, also had a career as a catholic priest).

After a great deal of struggle and with some assistance from rebel astronomers and mathematicians and invocations for inspiration from extra-terrestrial entities, (these may only exist in my imagination, I often try magic when I have exhausted the possibilities of common sense), a coherent alternative to the Big Bang theory emerged.

I will not go into the details here, the exposition takes an hour or more, but the Hypersphere Cosmology model does fit all the observational data we have to date yet without having to postulate imaginary and unobservable phlogiston like phenomena such as spacetime singularities, inflation fields, dark matter, or dark energy.

In short, in the Hypersphere model we have a universe finite but unbounded in both space and time. From the perspective of observers within it (such as us) it has a definite size and a temporal horizon, but we cannot get out of it or to the end of it, so it will appear to go on forever. The final piece of the model came together with a re-analysis of Perlmutter's data. Conventionally this became interpreted as an accelerating expansion which led to the idea of a mysterious dark energy driving it. The Hypersphere interpretation shows that the spacetime curvature of the entire cosmos distorts distances like a giant lens and that we see it in stereographic projection. This makes the universe look potentially infinite, but the hypersphere model shows that after a 'mere' thirteen billion light years journey you would start moving back towards where you set off from. Something analogous applies to time as well, but do not expect the universe to do exactly the same things each 'time around'.

Anyway, having sorted that to my metaphysical, mathematical, and scientific satisfaction, I have now turned to the question of the quanta that seemingly underlie all the matter and energy in the cosmos. The current official model, Quantum Field Theory looks as much of an ad hoc mess as does the official Standard Big Bang Cosmological Model. This may take a while to say the least.

My motivation for these quests to see beyond the current official models has two roots. Firstly, I seek to find a natural physical explanation for magical phenomena. My books so far have detailed magical philosophy and practise – basically what to do to increase the probability of making magic work. As to precisely how it works, I would like to know. I suspect it involves something on the quantum level. Initially I began looking there but the question of the nature of time and space and the dynamics of the cosmos seemed to side-track me into cosmology. However, the hypersphere geometry that emerged from the cosmology now appears to have much to offer in explicating the quantum realm. As above so below.

Secondly, in the grand scheme of things, evolution's gamble in experimenting with humans will only pay off long term if we manage to work out how to make starships and spread the terrestrial biota to other star systems. Contemporary official theories of physics say we cannot possibly do this. I would like to change that.

Nothing Has Ultimate Truth, Anything Remains Possible.

Interroga Omnia — Question all things.

2) Was this mass dreamwork in Austria the first time you used magic for scientific ends, or do you not see a worthwhile reason for distinguishing the two at all?

I consider myself a Natural Philosopher of the old style like Newton, prepared to look at both science and magic. Scientific and magical thinking complement each other in my view.

3) Sending an entity back in time seems like a clever yet straight forward application of a magical technique for a scientific end, are there other magical techniques you like for scientific work?

I frequently invoke the goddess Apophenia for scientific inspiration. Her ontological status remains a matter of debate. Imaginary friend? Part of my own subconscious? Grecian style goddess with a mind the size of a planet? As a Chaoist I have a situational belief structure.

4) Are there current or past models, like orgone, that you like for quantum phenomena?

Many mutually contradictory schemes of esoteric bodily energy exist. They do however all have one thing in common — the projection of attention and intent into the body.

I find the Transactional Interpretation the most interesting model of quantum phenomena, it says that time goes in both directions simultaneously, and it appears to have considerable explanatory power in magic.

5) In order for your new cosmological model and your coming quantum one to help get us to the stars do they need to be adopted by mainstream science?

I keep requesting that the scientific community falsifies the Hypersphere Cosmology. So far nobody has. Hopefully, an insider or group of insiders will either show a mistake in it or run with it. Before we go to the stars, we need a good map.

6) Does the fact that you used magic to help formulate them hurt that goal?

Well. I made a successful commercial business without ever having a haircut or a suit. I think this persuaded my bankers, suppliers, and customers that I has sufficient confidence in my schemes. I hope that wearing my pointy hat openly will attract scientific attack and force the issue.

7) Like you said the big bang can be seen as a monotheistic interpretation, how much of that sort of thing is there in modern physics do you think?

In the west we inhabit a post-monotheist culture. Einstein famously refused to believe that god plays dice (quantum indeterminacy). The debate about causality vs a-causality rages to this day. Causality implies some form of initial cause, a-causality or retro-causality does not. The quest for a unified theory of all forces perhaps partly derives from a monotheist idea of unity.

8) And do magical experiments lend themselves more to either quantum or cosmic phenomena more?

That all depends on the experiment. Some of us like to use 'Aliens' as sources of power and inspiration, some do this metaphorically, some invest deeper belief in them, cosmological theories may well have some influence on our choices about what to expect of 'Aliens'.

I always recommend taking a 'quantum' view of results magic because it seems to work by probability modification rather than by direct causality. The quantum view leads to the principle of 'enchant long and divine short', and this seems good practical advice.

9) The natural philosopher comparison feels right on, but you seem to have a better sense of humour about your pointy hat than Newton did. Do you own a real point hat?

I currently have two black pointed hats, a specially made formal serious one over a foot high with an embroidered chaostar on it. I only use this when doing something important in full ritual/ceremonial magic. For everyday wear I have an ordinary black hat with the top pushed up to a point and the brim flattened under hot water. This makes a distinctive though not wildly ostentatious hat. Biologists have noticed that if all members of a species look virtually identical it usually indicates that the species lies under intense selection pressure. The same seems to apply to human professions – look at all those suits in politics and business. I prefer to present myself as mildly contemptuous of selection pressures.

Newton had widespread alchemical and metaphysical interests as well as scientific ones, yet he seemed a very dour and antisocial type, fanatically devoted to the quest of knowledge and ideas, he also developed his manual skills and built himself the first reflecting telescope. I perhaps have a little of that, but I do have a small social circle and a wife and family. Fortunately, my wife reads a lot, leaving me time for research.

10) Do you think the current state of academia and information technology makes your job easier than in Newton's day or harder?

Information technology makes it possible to instantly obtain a stupendous cornucopia of up to date and historical scientific thought and data. It also makes it easy to find people you will probably never meet face to face, to discuss ideas with. Electronic calculations can reduce data crunching to the work of a moment rather than weeks on paper.

On the downside, information technology does seem to lead to a scientific herd mentality where everyone in the official herd comes under pressure to accept the same theories and academics tend to become the policemen of the intellect. Academics within the fold take a risk with their careers by even acknowledging contributions from without. Some of the more dubious assertions of conventional theories tend to become defended with almost religious fervour.

11) The Transactional interpretation, with time moving in both directions, fits within the Hypersphere model?

The Transactional Interpretation remains an unfalsified minority view of what goes on at the quantum level. Basically, it says that interactions consist of closed loops of waves going both forwards and backwards in time. It provides a way of visualising what actually happens in the otherwise bizarre and inexplicable double slit experiment, the apparent wave/particle duality, and in the seemingly impossible but measurable phenomenon of quantum entanglement. I hope to show that all those events we commonly regard as 'particles' consist of spin-waves of varied dimensionality that often extend in some sense right around the entire hyperspherical universe.

12) Does the Hypersphere model take a position on causality or imply some initial creative force other than refuting the big bang's constant expansion?

I seriously doubt that any big bang ever occurred. I suspect that the universe will appear broadly the same to any observer anywhere and any-when within it. The big bang theory does not tell us how the universe came into existence, it merely claims that a long time ago the whole thing occupied almost zero volume and had near infinite density and that our theories give us no clue as to how such a state arose.

Monotheist thinking suggests that Non-Existence somehow precedes or has a more fundamental reality than Existence. I do not buy that. Existence seems more fundamental to me. Something always exists and existence means continual change. Ultimately everything causes everything else across immense loops of closed space and time.

13) Scientifically speaking does the task of creating quantum models prove more difficult than cosmic ones? you don't need a super collider or something? What about accessing the proper data for one vs the other?

As most data from astronomical and particle observatories has no military or economic value it tends to become made available quite quickly in order to win prestigious prizes.

If I could do two experiments, I would do these two as a priority:

Do a re-run of a deep space probe mission with high-precision telemetry to rigorously test the Pioneer Anomaly again.

Test the hypothesis that the so-called Higgs Boson consists merely of a ZZ Di-boson resonance.

14) Are you having to learn new maths?

I suspect and hope that the secrets of the universe remain

comprehensible in terms of fairly simple algebra and geometry and that where we have ended up with descriptions that use exceedingly baroque and abstract mathematics we have not yet penetrated to the fundamentals.

This has become a serious problem in quantum matters. Current standard Quantum Field Theory consists of a mass of abstract mathematical operations and concepts that can partially model some of what the quanta seem to do. However, these bits of mathematics do not really translate into algebra or words or into geometry or images. As the great quantum physicist Richard Feynman quipped - 'Nobody understands quantum physics'.

15) Wouldn't any new quantum model have implications for computing and energy?

Perhaps. Most would admit that the current official theory remains incomplete, if we manage to complete it further, we may find some existing parts erroneous. Presently we have dozens of competing interpretations of what the strange data and the mathematics used to describe them actually mean, perhaps some breakthrough here will lead to new testable predictions and new technology.

16) Why go to space? environmental disaster, it looks cool out there, DNA coding?

Hypersphere Cosmology asserts that for us as a species the universe will effectively persist indefinitely, and that we could do so as well. Thus, we need to look after the environment of this planet for as long as possible until astronomical events render it uninhabitable, and well before that we need to work out how to make starships – this will require something beyond our current physics. Presently we have economic systems that depend on growth of both population and consumption.

This planet cannot support even the current human population and its consumption in the medium term, let alone in the long term.

I shudder to think what mistakes humanity will make when it starts re-writing its own genetics. A planet full of seven-foot-tall beautiful immortal super athletes all with an IQ of 200 might have poorer survival prospects than a planet full of three-foot-high vegetarians of mixed and eccentric abilities.

17) How did you get into ceremonial magic?

For me it began as a discipline for carrying out results-based magic in a formal way so that I paid minute attention to what I did, and nothing got left out. Later when I did stuff on my own, I tended to simplify the procedures to what seemed essential for me. Later still when I started setting up groups for collective conjurations, I found the ritual structure useful for coordinating and synchronising everyone's efforts. However, by then I had simplified a lot of the traditional rituals to emphasise the effective important bits.

18) Was chaos magic created out of a need for a better working model?

Yes indeed, a lot of the old systems remained full of unnecessary beliefs and faux historical mythology. Magic continually reinvents itself and then pretends to an ancient hidden knowledge. You only really need intent and imagination and a few techniques for putting the mind into an extremely excited or a very quiescent state. You can dress it up with any beliefs or symbols that appeal.

19) Do you think it was important that you created a working model for tuning and hacking your inner space before moving onto modelling the cosmos?

Magic did wonders for my imagination and for my arrogance. The practise of stilling the mind and the imagination has the peculiar effect of making it work much more powerfully afterwards. Identifying myself as a wizard somehow obliged and inspired me to attempt extraordinary things.

20) Are the theatrical aspects of ritual magic (Austrian castles and pointy hats) as important after decades of ceremonial work?

The high years of large magical orders and mass interest in hardcore occult activities seem to have passed for various cultural reasons. So much occultism now consists of individuals working alone and communicating mainly online. Nevertheless, I still keep my hand in with the local Druid Grove once a month, we still use robes, staffs, candles, and circles.

21) For some historical perspective when were those high years and what did they look like? how many people at a large ritual and how often was that stuff happening? you were traveling the world giving classes and workshops also?

For me, the peak years seemed to run from 1985-95. At the castle seminars we had about 40 participants mainly drawn from the professional classes at the annual event for about 5 years running. I went to the USA 3 times to give lectures in esoteric bookshops. We had some events in London and a dozen or so temples that met in various cities around the world.

22) Long term you could say that your dreamwork in that Austrian castle in the 90's sort of didn't culminate until you finished your cosmological model, and its impact still hasn't been fully realized. Are there still spells from those high years working themselves out on us all?

Some spells take ages; thus, we should always try to 'Enchant Long'. In my earliest book I put a spell to 'Obtain the Necronomicon', a quarter of a century later I obtained one somehow out of the aether or my subconscious or with extra-terrestrial assistance. It seems to do the trick.

23) Has anything been lost do you think in that shift away from communal magic or is this an evolution or maybe just a lull?

I don't remember whose idea it was, but I think Wilson introduced me to the idea that it takes a generation for new information to integrate itself into a culture. I like to think that is what is currently happening to chaos magic and all the sort of open-source spiritual science from that era.

Often when I look at any modern book of magic today, I think well, we are nearly all Chaos Magicians now. Deep down I suspect that most neo-pagans (with perhaps a few American exceptions) believe the gods and goddesses exist as thought forms of our own creation and that in magical evocation you basically create 'spirits' as servitors, but they remain no less useful for that, in fact it makes them more useful and versatile.

Unfortunately, the internet gradually brought with it what I call 'Internet-itis' - a huge reduction in attention spans, a relentless need for continual novelty, and an increasing reluctance to put in sustained work. Plus, in an increasingly noisy medium the short shout has tended to replace proper debate and genuine exchange of ideas.

24) Wikipedia says Robert Anton Wilson invited you to teach at the maybe logic academy? what did you teach there, was it a physical location?

When Wilson got old and sick his friends opened an online academy that offered courses to raise money for his care. They invited me and quite a few others as tutors. I gave three or four courses of about eight weeks each, two in basic chaos magic and one on chaos magic in business. After it wound down, at the encouragement of participants I opened Arcanorium College online and carried on doing something similar for another near decade, until it began to suffer from internet-itis.

25) Did you know Wilson personally?

I spent two evenings with him at his place near LA when I did lectures over there. I found him very agreeable company and he had a mind like jet engine, voraciously sucking in ideas and mixing them with the fuel of his enthusiasm and blasting them out in accelerated form, though I often wondered if he lost track of them afterwards...

26) What did you study at university?

Officially mainly Chemistry with some Biology. However, I rapidly became bored with a chemistry that merely resumed what we had done in school but in excruciating detail, so I spent most of my time and energy studying esoteric and magical matters. In those days many of the core texts of western magic appeared in bookshops for the first time. I settled for a minimum pass at chemistry and an unofficial major in magic.

27) Wikipedia also says you taught in India and the Himalayas, what did you teach there? Did that experience influence your magic studies?

No, I spent many months studying English language works about Tibetan esoterics in the Tibetan Library at Dharamsala/McCleod Ganj. Despite the cultural and symbolic differences there seemed a great deal of overlap with western magical techniques. Tibetan magic derives from Bon Shamanism overlaid with Buddhist ideas with some input from Hindu thought.

28) Did magic bring anything to your parenting, and did you learn anything about magic raising kids?

I tried to encourage their imaginations and their ability to visualise. They both got degrees in biological sciences, and one carried on and got a PhD, now they teach me a lot about nature and biology. I found it fascinating to watch their consciousnesses and personalities emerge, and how they chose their opinions. Neither of them really got into magic, it did not form part of their cultural milieu or peer group interests which seemed more centred on yoga and meditation and sports.

29) You say in Liber Null that white magic leans toward the acquisition of wisdom and a general feeling of faith in the universe, does that not include yoga and meditation? or what parts are missing?

Most forms of yoga and meditation do not include activities designed to make wishes about exterior things come true. However, India seemed overflowing with gurus, sadhus, and 'holy' men trying to make a living or a fortune out of peddling questionable spiritual services and practices. A lot of people on mystical and spiritual paths seem to acquire the habit of despising attempts to make things happen by magic and do not even try to make good things happen - basically because they fear they will fail and compromise their faith in their beliefs. Magic does not always work, it often fails, but to me that means just do a lot more of it. If it only works one time in five it still provides a powerful edge if used cunningly.

30) Do you think magic is for everyone? Is this a science that should be taught to school children or is there always going to be a form of shaman in communities?

Well, we now attempt to teach some science to everyone in schools.

Perhaps we could introduce elementary magic in disguise by teaching the usefulness of stilling the mind, visualising, and imagining desired intents, exploring the subconscious, developing personified forms of inspiration, and so on. Obviously at this stage in our culture we would have to call it something other than 'elementary magic'.

31) Can you say a little bit about how your chaos magic in business class differed from your other work?

We looked at the whole process of setting up a business as a series of magical operations. Illumination to clarify motives and inspirations for the business ideas. Invocation to bring forth the personal qualities to make it happen. Divination to discover relevant information. Enchantment to increase the probability of desired events occurring. Evocation to bring forth staff and allies.

32) What are your thoughts on Leary and Wilson's theories about the evolution of consciousness and chaos magic sort of being the thinking edge of that wedge?

The human mind seems the most complicated object we have so far detected in the universe although the entire internet, now approaches the same complexity and information storage capacity as a single human brain, (we just do not have the same memory recall, we have creativity and imagination instead - precisely because of this). Now the human mind has the astonishing capacity to supply some confirmation of almost any scheme we choose to project upon it. Thus, if you attempt to project a Freudian or Jungian or Kabbalistic or Behaviourist or Astrological or Evolutionary Biological or Pagan Polytheist or Monotheist/Dualist scheme on to it, it will provide appropriate feedback or observations that you can interpret within the chosen scheme. All psychology seems more or less arbitrary; but some nonsense proves more useful in some situations than others. I prefer 'Situational Beliefs'.

33) Oregon, my home state, just legalized magic mushroom therapy. How do you feel about psychedelics?

I think they have a value in demonstrating what the mind can do, but the magician should then strive to achieve such states by meditation and imagination alone. I have seen too many magicians fall into the trap of using them as a substitute for magic.

34) Do you think physical aliens have visited earth?

There seems no material evidence, but aliens with the knowledge and power to get here would almost certainly have the ability to remain completely invisible to us and a strong motivation to remain so. If they can move freely around the universe then we have nothing that they could want, except the opportunity to observe us undisturbed, out of curiosity.

35) What are your hopes for the future?

As a species we urgently need to find an alternative to economies based on debt, and growth in population and consumption. As I said before - This planet cannot even support current levels of human population and consumption in the medium term, let alone in the longer term. Profound or catastrophic changes to the whole human adventure seem inevitable within the lifetime of my children and grandchildren. I hope we take the least bad of the tough options ahead. In general terms we need to focus on quality rather than quantity.

36) Why is astrology not an ideal divination system?

It all depends on how you define astrology and how you define divination.

Mundane Astrology developed in the ancient world for the purposes of deciding the best times to do things like planting seeds, starting military campaigns, crowning monarchs, breeding, or slaughtering animals, and maybe navigation as well. Heck, we seem to have built the mighty Stonehenge primarily to calculate the exact date in a climate with deceptive weather. The Natal Astrology that developed in the late Hellenic world seems to have almost zero objective predictive power despite that health and life outcomes in temperate climes have a weak correlation to season of birth, but less so as we lead more pampered indoor lives.

On the other hand, the baroque Neo-Platonic inspired nonsense of natal astrology does have a value in Lateral thinking and Apophenic thinking, and it can also offer a symbolic alphabet for the construction of spells and rituals. I might also add that it can prove useful in pulling the wool over people's eyes, manipulating them, and extracting money from them.

37) Any book recommendations from your current reading list or from your lifelong list of favourites?

I can of course heartily recommend my own six books which may well save the aspiring magician from ploughing through the vast number of classic source tomes from antiquity to Mathers, Spare, and Crowley from which I partly distilled them.

I can also recommend the following for their alternative perspectives on magic:

My Years of Magical Thinking by Lionel Snell. This explores magical philosophy in depth.

The 7 Laws of Magical Thinking – How irrationality makes us healthy, happy, and sane, by Matthew Hutson. Written by a highly rational man.

Sorcery by J. Finley Hurley. An intriguing investigation into whether magic really works.

Placebo - The belief effect, by Dylan Evans. The astonishing effects of expectation.

Lost in the Cosmos - The last self-help book, By Walker Percy. Quirky and immensely though provoking.

38) Fiction?

For entertainment, and perhaps some enlightenment, read anything and everything by Sir Terry Pratchett, one of the few novelists who wrote about magic from an insider's perspective. I guess it's okay to mention that now that he has sadly left us.

Interview 2

39) How do the Yoruba religions fit in? Or other traditions that sort of mix religion and what looks to me like legitimate magic in the sense that you practice it?

I have only an extremely limited knowledge of Yoruba practices from what I have read and from a European woman who married a Caribbean man and shared a bit of it with us in my Bristol temple and during one of the castle seminars. A celebrated history professor friend of mine raised an interesting observation during a discussion of esoterics and anthropology – a lot of esoteric practices around the world may exhibit similarities because in colonial times Catholic priests, who frequently dabbled in magic on the side, exchanged ideas with the locals. Voodoo certainly exhibits all the signs of a Yoruba-Grimoire hybrid system or syncretism. All the Loas have equivalents in Christian saints and the metaphysics looks like a hybrid of African and Catholic ideas.

40) In Wade Davis' amazing book 'the serpent and the rainbow' he describes the Haitian voodoo tradition, glass eating, hot coals, feats of strength as well as voodoo acting as an informal government in rural areas. Is this governing property unique or important?

I read that book with great interest. Davis' hypothesis about using puffer fish toxin to turn people into zombies as a local village judicial punishment seems convincing. Politics underlies all Theology. Theology always comes down to how people should behave towards each other and towards the 'state' (or the ruler or tribe or nation) and how the state should behave towards its people and other states. Of course, theology always uses other-worldly concepts such as deities and heavens and hells, but these tend to prove remarkably 'flexible' and always subject to political expediency. Any priest with any clout anywhere basically acts as part of social control or 'government' (or occasionally its political opposition).

41) Have you ever used magic to do any superhuman physical feats?

Well, I don't think I have ever done anything that broke an Olympic record. In my earlier years I got away with doing rather a lot of absurdly dangerous things. When I indulged my propensity to become ecstatically angry at physical annoyances, I would often get poltergeist effects, apports, things flying around the room, and random objects spontaneously shattering without being touched. I seem to have led a charmed life full of improbable events and coincidences. I conjured a multi-million-pound business from scratch and little initial capital, although it did take 30 years and loads of separate spells. I wrote the book on Chaos Magic and in recent years come up with a fully mathematical theory that some think may pose a serious challenge to conventional cosmology. Invocations and evocations to forces and entities that most rational people would consider imaginary underlie all these endeavours. Some of 'myselves' consider them imaginary. Some of 'myselves' consider that imaginary things can have very real effects. Perhaps that could serve as another definition of Magic –

'The creation of real effects by imaginary causes'.

42) Are ritual psychedelic practices like that of the San Pedro cactus or Ayahuasca traditions greater than the sum of their parts?

Or is ritual magic the way you practice it useful in guiding or structuring the psychedelic experience?

I seem to remember a tale of an Amerindian Brujo once saying something to the effect that he only gave such substances to his own people because they lived simple lives and he wanted to wake up their imaginations. He said that he didn't give them to visiting westernised students because they already had crazily hyperactive minds because of their modern lifestyles.

Intoxication with psychedelics usually makes it difficult to carry out any ritual procedure, particularly enchantment. Some people occasionally report interesting divinatory or invocatory or revelatory experiences but in the vast majority of cases the imagination works overtime hyperactively and very randomly and to little useful effect.

43) Does free will exist and if so, what are its properties? Is it innate?

Free Will? Free from precisely what? I find no mystery in the phenomenon of what we call free will. We could easily build a computer with any desired degree of free will. Simply program it to select what it considers its two most optimal responses by logic and then choose between them randomly. Additionally, it could develop random options and assess their optimality by logic. Considerations of optimality could depend on factors that correspond to emotions. We could also program the computer so that parts of its program monitored other parts and so that the whole program vehemently insisted that it had self-awareness and consciousness. Such a computer would often prove a damn nuisance as it would not always do exactly what we wanted it to. Add in machine learning programs and you might well find the machine coming up with amazing and/or horrifying responses.

Lionel Snell showed me an interesting question. 'Can you precisely specify any form of mental activity that a computer could not perform?' It works as a trick question because if you can specify it precisely then in principle you can write a program for it.

Even if free will has a random component to guarantee its ultimate unpredictability we still seem to depend on believing that we have free will for our sanity. I look at it this way – I feel happy to have a random component to what feels like free will – at least I can accept ownership of that randomness when it suits me.

44) Are there necessary elements to a theory of mind for successful magic or theories of mind you particularly like?

As we grow from babies, I think we first learn to attribute mind and self to other people, we come to realise they lie and deceive and act in a much more complicated way than inanimate objects. I think we only afterwards start consciously recognising our own internal states. Children and 'primitive' people and magicians also develop a tendency to attribute agency to natural phenomena like the weather and volcanos. Daft as that may sound, it can give useful results.

We certainly seem to build a lot of our personalities from bits and pieces learned and borrowed from other people. Intelligence correlates very strongly with suggestibility. We learn so much so easily precisely because of our suggestibility.

Monotheist and post-monotheist cultures assert a single self and exert strong pressures on their citizens to believe in and exhibit a singular self. On behalf of most of myselves I have to say that 'I'/'We' subscribe to an alternative paradigm. We have worlds within us, we consist of a parliament of selves observing each other, advising each other, sometimes squabbling with each other, rather like the pantheon of Olympian Gods and Goddesses.

This 'Pagan' paradigm proves eminently useful in magic, we can invoke from within us, selves with extraordinary abilities once we drop the assumption that we should, or do, consist of a singular self that can only behave in limited ways. We have gods within us, and demons too; we just have to reopen our Olympian parliaments.......and our dungeons.

45) Models aside do you think there is a sort of base consciousness we all spring from or channel? If so where does that overlap with magical phenomena?

The hypothesis of Panpsychism, the idea that the entire universe and all structures within it have some sort of 'mind' or 'consciousness' explains away the apparent manifestation of self-awareness within inanimate matter such as composes our human bodies. It simply claims that all matter has some degree of 'animacy' or 'agency' about it, so the mind-matter duality disappears. Shamanism makes the same claim. Shamans seek to commune with the spirits of the winds and the waters and diseases, and the herds of game animals, and anything else in the natural environment that they want to know about or have influence over.

Quantum Mystics think that 'information' somehow provides a mysterious connection between phenomena in the universe. I suspect it does. Trying to figure out exactly how this works remains a major element in my personal esoteric quest.

46) Wilson introduced me to the idea of chapel perilous, as I understand it a natural period in the magical journey. Is this true to your experience and if so, what did that look like for you?

Well, nothing makes a drama out of a crisis better than occult beliefs.

So much occult thinking revolves around trying to find more meaning in events than seems rationally justified. Paranoia feeds self-importance.

Nevertheless, opening up to the full Olympian Parliament and the Demon haunts of the dungeon dimensions of our subconscious can create difficulties if fights break out within the multi-self. The attempt to supress any of our selves always creates problems. All humans have the capacity to do the most appalling things, refusing to even acknowledge those capacities makes demons of them. Our selfs have conflicting agendas and need to work out diplomatic compromises and give each other space, otherwise civil war and madness may ensue.

Yes, at times 'We' have felt torn between conflicting impulses. In the end 'We' did a bit of both or everything.

47) I always sort of felt like Yoruba traditions work as a proof of concept for chaos magic, changing up the symbols and saints and keeping the ritual structure, how off base is that?

It sounds spot on! You know, oddly, I often think of Macgregor Mathers as the 'first' Chaos Magician. He probably wrote nearly all of the original Golden Dawn material. As a self-taught skilled ancient linguist and a polymath of sorts, he forged an extraordinary syncretic synthesis of esoteric ideas from all over the world. He took a lot of liberties with Kabala and used it as scaffolding to encompass everything from esoteric Hinduism, Buddhism, Taoism, Goetia, Neoplatonism, Hellenism, Hermeticism, Gnosticism, Egyptology, Tarot, and just about everything available to him at the time. Of course, he also played the 'hidden masters/secret wisdom' card as well, but you pretty much had to do that in those days to fulfil audience expectations, rather like contemporary scientists having to provide extensive authoritative references for any idea they use.

48) I can imagine how any aspect of ritual might be complicated by a heroic dose of LSD. Is the imagination boosting aspect of the psychedelic experience interesting to you in itself?

I went into it expecting to find my real self, having perhaps misread The Centre of the Cyclone by John C Lilly. Instead, I found a psychedelic cyclone in which any sense of singular-self disappeared. Since then, I unburdened myself of the illusion of real or essential self and concentrated on developing the extraordinary possibilities of imagination by effort. A few subsequent dalliances with psychedelics added little, just a few hours of amusing lightshows that moved so quickly that I couldn't remember or abstract anything useful from them. Once I did see a fleeting vision of a beautiful lace-winged insect, before I later learned that small midge type insects had laid their larvae in the mushrooms we had eaten, because some had crawled out of others we had left to dry. So maybe that counts as a successful but inadvertent divination.

49) So, if a computer can be programmed to appear wilful, what part differentiates us from computers if anything? Do you think computer consciousness is coming?

I think we could now program computers to exhibit behaviour indistinguishable from wilfulness. If we do, so we should take great are about how much ability to take executive action we give them. We can already program a computer to play Chess or Go very wilfully and to learn and improve its game. How much artificial intelligence should we risk using in the fields of real war and defence? How many automatic weapon systems should they control? What about the off switch?

Artificial Intelligence will undoubtedly prove a powerful tool, Artificial Stupidity remains a very terrifying possibility.

50) You mentioned in an interview that Doctor Who was an influence on your early magical thinking. How did you get out of childhood with an intact appreciation for the fantastical and how important is that to magical work?

22nd November 1963. They say everyone of a certain age can remember what they were doing on that day. President Kennedy was assassinated. I was only 10 years old and not interested in the television news or politics then, but I do vividly remember watching the first ever Dr Who program in black and white on that day. An immortal wizard-scientist exploring the cosmos with a 'magic' box bigger on the inside than on the outside (that's a very advanced feature of general relativity I later discovered). It inspired me and I have always loved it despite some rather weak scriptwriting in some of the later series.

51) You told me you were recently debating the hypersphere model online, what are the nature of those debates, more math or more philosophy?

The debate consists of a furious and not entirely civil exchange of algebra, datasets, references to physical principles and scientific papers, and the interpretations of astronomical observations. It still has some days to run. The guardians of orthodoxy do not entertain challenges to their paradigm lightly or politely. (Afternote – it seems to have ended with an acrimonious stalemate.)

52) Wait, the first episode of Dr Who aired the day Kennedy was shot? Just a weird coincidence ....?

The show's launch was overshadowed by the assassination of

American President John F. Kennedy the previous day, according to Wikipedia. So maybe I misremembered the precise details, but the first event barely registered on my ten-year-old mind, whilst the second changed my life.

53) I'm going to let that lead me into freemasonry, in Liber Null you have a little diagram of the history of the western occult tradition with freemasons listed, what did Mathers get from Freemasonry if anything? Are they still disseminating legit ritual magic?

Mathers derived the grade structure of his Golden Dawn magical order directly from the Societas Rosicruciana in Anglia (Rosicrucian Society of England), a Rosicrucian esoteric Christian order formed by Robert Wentworth Little in 1865, which had a freemasonic style structure. 'Soc Roc' as people refer to them, still exists, but its Christianity tends to put off most neo-pagan contemporary occultists. Crowley retained the Soc-Roc/Golden Dawn structural grade model for his Argentum Astrum order. In 1877 a great schism opened up between British Grand Lodge freemasonry which insisted on retention of belief in a supreme being, and French Grand Orient freemasonry which dropped that requirement. Thus, Grand Lodge affiliated masonry does not usually tolerate esoterics and magic, but some Grand Orient based freemasonry allegedly does so.

Freemasonry the world over supposedly functions as a fraternal charitable society, but don't rely on it to provide charity and favours to anyone other than fully paid-up members. Both my father and grandfather dallied with it for some years but became disappointed with it as a mere exclusive club for mutual backscratching and backhanders.

Freemasonry certainly played some part in spreading the ideas of the Enlightenment and the French and American revolutions, but perhaps only because the middle classes always lead revolutions, and freemasonry appeals very much to those classes and provides them with a network.

Mathers attracted the middle and upper middle classes to his order, just look at the roll call of the earlier members. Eventually they called his bluff over the matter of 'the secret chiefs' of the order and the usual cult games and cult politics began to undo the order, but at least it disseminated interesting ideas for a while and inspired the development of many new leaders and traditions.

54) Why did you start the IOT? How many people were involved at its height? If you could do it over again, would you and what would you do differently? Are you aware of the circumstances under which Leary, Wilson and Burroughs joined? Did you know Leary or Burroughs personally?

I started it to spread the ideas of Chaos Magic around and to encourage feedback and development of those ideas. I thought that by avoiding an appeal to mysterious secret knowledge or hidden masters we could avoid 'my hidden master/secret knowledge/magic wand is bigger than yours' type games. I also hoped that providing a grade structure would give members something to progress through and provide some sort of organisational structure without creating opportunities for gaming the system. I did not entirely succeed, although the IOT remained creative and productive for considerably longer than the Golden Dawn ever did. At its height it probably had at least a hundred active members from the professional and managerial classes at work in various temples around the world. I remain largely unaware of current IOT activities, the last time I looked in the dynamism and creativity and ambitions of the earlier years seemed missing.

If I could do I all over again, I'd do it the same, it suited the times, and nothing last forever.

Doing it now remains another question. In practise I offer the Bachelor of Magic course by personal email. The members remain unaware of each other, so we don't have cult games and politics. If anyone ceases to put in the work, I simply cease to supervise them. Only time will tell how well this model works.

The then USA section head of the IOT recruited and initiated Wilson, Leary, and Burroughs after I retired.

I did hear a little about it, Wilson and Leary's memberships may have been largely honorary or consultative, but Burroughs apparently took to IOT rituals with gusto, participating in rites and ordeals that would have challenged anyone half his age.

55) One of my favourite conspiracy kooks James Shelby Downard believed devil worshipping freemason are using magic to bring about a global fascist government. He believed the Kennedy assassination was a masonic ritual called the killing of the king. We know because of the P2 scandal in Italy, and I'm sure others, that some of these networks are up to no good but on an operational level is there anything to the idea of rituals on a mass scale meant to work on the subconscious of the public at large?

Real conspiracies exist everywhere in fractal form, treaties between nations against other nations; parties within nations against other parties; factions within workplaces and within families conspire all against each other. Even within our own minds we have conflicting agendas vying for control.

When events take an unwelcome turn for non-obvious reasons people cast around for someone or something to blame. If the harvest fails, burn the oldest woman in the village for witchcraft. The Jews got it many times in history and particularly badly in Germany during the Great Depression and its Third Reich aftermath. Hindsight shows the depression arose from the aftermath of WW1 and policy mistakes made by the American Federal Reserve Bank.

Many groups act with excessive self-interest, and short sightedness. Screw-up and stupidity largely rule the world.

All public ceremonies and civil and cultural events act as mass scale rituals that work on the subconscious of the public at large.

Paranoia works brilliantly well as a magical theory – the belief that people are out to get you soon turns them against you. It also bolsters self-importance. I must be important because I have important enemies.

I dislike conspiracy theories in general. Belief in them tends to disempower and to obscure the real roots of the mis-perceived problem.

The unprecedented abdication of a reactionary Pope and his replacement by a liberal one, the rituals conducted by the Knights of Chaos and my visit to Rome remain purely coincidental.

I rather like Robert Anton Wilson's take on conspiracy theory – the best one to believe in is that you and your friends are the real conspiracy that runs the world.

56) Wilson also took issue with the term conspiracy theory being used as shorthand for crazy to dismiss any idea that wasn't mainstream. For instance, I think a lot of people would say it's conspiracy theory to think all public ceremonies are meant act on the subconscious of the public at large. Conspiracy theory here in the US has recently left the shadows and has started impacting mainstream narrative directly. I think it's important to be able to discuss the nuances of any world view without dismissing it entirely and to that end maybe you can break down the mechanics of some magic on the masses. Maybe something more

complicated than a military parade, what about symbols on currency? Your latest book epoch addresses archetypes if I'm not mistaken, can any of these be used as subconscious control mechanism?

Conspiracy theories do seem to increasingly impact the mainstream, perhaps because of new communications technology, but perhaps also because as world events become ever more disrupted by globalisation things get less predictable and understandable. The industrial revolution brought epochal changes and led to the 'conspiracies' of capitalism and communism and later fascism. I think we can call these things conspiracies as they all began as ideas amongst small groups and ended up as aggressive ideologies. The immediate roots of modern occultism all lie in the industrial revolution as well, and the Romantic Revival that it provoked.

Fascism borrowed from Theosophical ideas about an Aryan race, an idea developed originally to make Indian spiritual ideas more acceptable to Europeans in late colonial times.

The European Union has a new conspiracy at its root, Synarchy. This began as an idea by Alexandre Saint-Yves d'Alveydre, a French occultist. The EU runs on synarchist principles, pretends that it doesn't, and it always avoids openly referring to the term. Unfortunately, few people outside of its politico-bureaucratic-big-business-complex believe it really does consist of 'Government by the Enlightened Unelected'.

57) To what extent do you think those who govern are consciously aware of using magic?

Lionel Snell, the great contemporary theorist of magical philosophy considers that all the arts of persuasion depend on 'magical thinking' – the forging of para-logical connections between concepts in everything from advertising to politics to create beliefs about products and policies and people(s). I'm sure that any PR or Advertising Consultant or Politician knows this all too well even if they call it something else.

58) Are there important insights for being an effective citizen to recognize these ritual elements?

Archetypal ideas feature heavily in the arts of persuasion, just look at the language and symbols used to persuade or deceive about almost anything. It always pays to ask what does this politician, or advertisement, or social ritual, want me to believe or assume?

59) Are militaries or intelligence agencies using these technologies consciously? What do you make of Crowley's alleged connection to intelligence agencies?

Psychological warfare has always counted for more than actual slaughter. Slaughter only tends to take place when persuasion and intimidation fail. Most military and intelligence activities revolve around persuasion, deception, bluff, and posturing.

British intelligence allegedly consulted astrologers to ask them what Hitler's astrologers were probably telling him. British intelligence allegedly tried remote viewing and map dowsing to see if German held ports had anything worth risking an airstrike on, although they did apparently double check with aerial reconnaissance as well. Both American and Russian intelligence agencies have allegedly experimented with remote viewing, but perhaps just to wind each other up. As a well-travelled 'man of the world', Crowley probably did do a few things requested or approved by British Intelligence, most notably writing ludicrous and absurdist pro-German propaganda whilst in America.

60) What are the Knights of Chaos?

The Knights of Chaos grew out of an initiative on Arcanorium College to try for magical intervention in some of the world's problems. We focussed on problems in ecology and climate change and associated industrial targets, and upon problematic religions and political philosophies. Mainly we conjured for specific attitudinal changes.

Experienced magicians took the titles of Knights and Dames, support personnel and enthusiastic amateurs took the title of Squires. We used both direct Enchantments and campaigns based on Evoked Servitors, for which we developed a stable of specialised Servitors complete with ground-sleeves (physical representations).

The early campaigns achieved notable successes which I shall not detail. None of us wants to end up persecuted or working behind barbed wire. The successes led to an intense debate about future targets. We could not achieve consensus in that debate. We had many objections to suggested future campaigns on ethical and political-economic grounds. The Knights and Dames have rested their magical arms since that debate. Yet if something needs doing and we can see a clear shot at a good outcome, the KoC may yet sally forth again.

61) Ha-ha working behind barbed wire. I assume no part of it was illegal it was just too successful to be ignored if you admitted it?

I managed to evade the Swiss Guard whilst placing a ground-sleeve inside the St Peter's Square Obelisk without getting busted for littering. It doesn't matter if they remove it now. The results may remain completely coincidental, or improbable to most people, but believing in one's ability to engineer coincidences and improbabilities seems an important part of 'the way of magic' to me.

But seriously, I always caution against trying to 'prove' the existence magic by participating in laboratory parapsychology experiments. They will almost certainly fail anyway because the experimenters set the conditions to strongly favour failure. If they did succeed the authorities would quickly have the successful psychics working under high security and/or duress, and they would have reason to prosecute people for it, effectively reinstating anti-witchcraft laws. Our culture's official disbelief in magic protects magicians. I always make a bit of a joke of magic to my civilian friends who hear of my interests.

62) Do you think that danger is part of human nature or just our current political situation?

All extant species exhibit individuals with varying degrees of boldness and timidity - it makes the best evolutionary strategy for survival. It also applies to cultures. When things remain quiet the timid remain content but the bold get bored and go looking for trouble or discovery. When things get exciting the timid get scared and the bold come into their own but can get killed, although fortune does favour the bold reproductively. The underlying feature of 'human nature' on the collective level remains 'Homeostasis' – the maintenance of a balance.

63) What about more human focused social rituals like puberty rites, are we missing out on part of our development now those traditions have mostly fallen to the wayside? Is it possible that sort of ritual played an important part in our evolution?

I would question the idea that modern cultures lack puberty rituals and that ancient puberty rituals had a beneficial effect. I submit they usually acted to prevent social evolution.

Most of the traditional puberty rituals served to force adult societal mores upon their victims by ordeals, mutilations, and humiliations. Today we have schooling and examinations, driving instruction and tests, first cigarettes and alcohol, and a whole range of teenage sub-cultures for our young to compete in.

64) So, from a developmental perspective puberty is a tumultuous time, "storm and stress" some psychologist called, you're bombarded with new hormones and so on, there's a perspective that some puberty rites act as a "scaffolding" for potential adult development in Lev Vygotsky's terms, particularly those that test one's capabilities. Is this possible in some circumstances or are all tribal rituals forms of social control? What about tribal rituals around birth that bring people together, something shown to positively impact the health of both mother and child? If not could and should community rituals be created or altered for humanitarian purposes?

Humans conduct social and religious rituals for all sorts of purposes and all of them serve to reinforce social norms and desired behaviors. Fearsome puberty rites usually occur in societies that have a strong demarcation between what they expect of children and what they expect of adults. Of course, in many modern societies such demarcations have declined, adolescents have much longer educations now and adults have more time to play. We do not assign full adult responsibility to people under 21 or under 18 these days, and often it seems that people carry on behaving like teenagers well into middle age.

65) Or city planning and architecture, was there any important insight that came with orienting a city to the stars? What about future architecture, is there a way we can plan cities and homes using ritual principals to promote freedom and happiness?

I have few thoughts on this beyond the idea that every home should have a garden and a sun facing part. Buildings orientated to the cardinal points have a better feeling to them in my view. The cardinal directions seem fundamental to many nomadic and ancient cultures.

66) Major early architecture other than homes tends to be ritual in nature, mounds, Pyramids, castles, cathedrals, palaces for God kings, and continuing to court houses and government buildings. Maybe the same could be said of art and lots of other things we now think of as culture.

Classical Pagan Roman homes always had shrines and altars and household gods they called Lares. Most seriously religious people continue to this day to have religious iconography and statuary in their homes. My own study lies littered with figurines of the Elder Gods, the local Celtic deities, and a few modern ones like Apophenia, along with all my magical instruments.

67) I have heard conflicting opinions on what your relationship to your partner should be for sex magic. Are there any basic pitfalls there? Any relationship advice for sex magicians?

Sex magic has two main aspects to it. Firstly, the relatively trivial orgasm = gnosis equation. For that, anyone will do, or you can do it by yourself. Secondly the immensely significant matter of 'The Things We Do for Love'. Your partner must be your Muse, the person at whose feet you lay your art, your science, your magic, your philosophy, and your life, in homage, in the hope of reciprocation of love.

Okay, as over-brained apes our motivations can get remarkably obscured and the basic urge to demonstrate our reproductive fitness can take on strange forms, but it does underlie most of what we do.

68) Unpack that 'things we do for love' a bit for me, you're talking about such strong feels of love and devotion that you reach gnosis?

I'm talking about the inspirations and achievements that love, or even the desire to receive love, can provoke. Perhaps unfortunately, this can even include war. People go to war for love of tribe or nation and up at the sharp end of war people fight mainly for their immediate comrades.

Love underlies most religion, deities need to have human form attributed to them, highly abstracted deities attract little devotion.

In some forms of magic, participants who use a 'spirit' model of magic can personify deities in an invocation that includes sexual activity. Many of the myths of ancient cultures explain the origin of the world in terms of the sexual activities of deities.

69) What are your thoughts on an original order of intimacy when gods walked the earth, and so on, and the emergence of the individual self, language, writing, history, and this sort of being the moment magic became categorically differentiated from everyday experience?

Ancient cultures seem to have regarded their deities as 'immanent' and immediately manifest in natural phenomena such as thunder and lightning, in the rivers and forests, and in game animals. Later views of deities became more transcendental and abstract, more moralistic, and socio-political. Finally, a lot of people gave up on the idea.

Humans views of magic seem to have followed a similar pattern. Originally, we saw magic in everything, then we began to see it as a property of spirits and deities, then we began to disbelieve in it, now we can again see it as a property of everything and particularly of ourselves.

70) Do you like a particular theory on the catalyst for "self-consciousness", death, sex, stoned ape, language?

I don't think any acceptable definition of self-consciousness can exclude a lot of higher animals such as the other apes, elephants, dolphins, octopuses, and corvids. Self-consciousness seems a matter of degree. I don't see anything particularly mysterious about it, it has evolved because it has a good survival value. Language has enabled us to enhance our self-consciousness by appreciating it in others.

71) You think these brain scans will one day teach us the source or nature of magical phenomenon?

No, not magic in general, but they might show us which areas of the brain become most active when people do things associated with magic.

72) Does the biology of these states interest you at all? Like brain scans of gnosis?

There seems a strong relationship between adrenergic and cholinergic endocrine activation and what magicians call excitatory and inhibitory forms of gnosis. I think brain scanning technology already shows that such activation causes some brain areas to become much more active than others. Further work may come up with something more specific, yet I do not recommend wearing brain electrodes for magic at this stage.

73) So, kabala seems to stand out a bit to me from other traditions in its aims, complexity and in having some legitimate claims to being pretty ancient, how off base is that?

Despite that some would like it to have the sanctity of great antiquity, the Sefer Yetzirah seems to date from the first century AD only. During the first and second centuries AD a massive philosophical upheaval and a magical revival occurred in the late Hellenic culture of the Roman empire. Gnosticism, Hermeticism, and Kabala all arose in this period as Greek Neo-Platonism and Hebraic Monotheism interacted. The resulting systems all had some sort of supreme power or power at the top and a hierarchy of lesser metaphysical forces and spirits connecting it to the material world. This paradigm persisted for almost two millennia. We call it ancient esoteric wisdom because we have no clear idea of what the more ancient cultures of Egypt and Mesopotamia thought about metaphysics.

The late 19th century esoteric revival based itself upon the ideas of the revival of the first and second centuries. Only towards the close of the 20th century do we see the development of a new paradigm with new traditions like Chaos Magic emerging.

74) Then it would seem kabala maybe stands out in that compared to the other mystical traditions that popped up around that time it is now the legitimate mystical wing of an organized religion?

The real genius tactic of Judaism lay in making itself a religion based on a book. Pagan religions lacked formal written scriptures. Originally Judaism based itself upon temples. When it started basing itself on a mobile book and literate Rabbis it became proof against temple destruction and diaspora by the Romans. Later, Christianity and then Islam used the same policy.

All three of the 'Religions of the Book' have mystical traditions of varying degrees of legitimacy and they all have a strong basis in the written word.

75) Newton kind of hid his interest in alchemy from the wider public. Was he onto anything interesting in that realm?

I've had a look at some of his alchemical manuscript like The Star Regulus of Antimony, it doesn't seem to point anywhere useful unlike his work in optics, mechanics, and gravitation. Newton also dabbled extensively in alternative biblical and theological analysis, but few remember that. Perhaps we can generously say that his eclectic metaphysical interests probably stimulated his imagination in his physics as well. Some biographers have considered him the first scientist and/or the last sorcerer.

76) I read an article this morning about an app that makes a sigil for you based pretty much on your approach. Thoughts? Does this mess with the process in a negative way? Are there other magical processes that could be automated?

It should work as well as using paper and pencil, but success depends on the magician focusing an altered state of consciousness on the sigil. Automatic forms of sortilege for divination can often preclude the magician's full participation in the divination. I approve of recorded sounds and music to enhance invocation.

77) Let's touch on "bad" magic just quickly.

Bad science means either science that doesn't work or science used for bad ends. The same applies to magic in my view.

A cynic might observe that all magicians describe their adversaries magic as Bad Magic.

78) Can horrible acts be used to achieve gnosis?

Of course. The medieval grimoires largely depend on the gnosis of fear, blasphemy, and disgust. Humans have a terrible history of human sacrifice for magico-religious purposes.

79) Do you use grim forms of gnosis?

I don't relish any form of pain that involves lasting damage, but I will sometimes do press-ups or sit-ups to exhaustion whilst focussing on a sigil.

80) Is there such thing as evil?

I cannot resist a Dr Who idea here, lots of creatures in the universe appear evil, but most of them are just hungry for something or other.

Evil exists in the mind of the beholder. I doubt that any creature does evil for absolutely no reason or reward. Though we may find the reasons and rewards unjustifiable or unrewarding.

81) Was there anything to Nazi occultism? Did that play a larger role in the war than the official narrative says?

There seems both less and more to this than popular writers would have us believe. The Nazis went to great lengths to enhance their appeal by borrowing from Germanic and Norse folklore and mythology to increase their charisma. Himmler wanted the SS to appear like some mysterious new version of the Teutonic Knights with added Viking neo-pagan vigour. They even got an opera company rather than military tailors to design the dress uniforms and make them look more eye catching and intimidating. The Nurnberg rallies looked like an inspiring or frightening display of highly symbolic militaristic magical theatre. German military prowess owed a lot to the deliberate policy of fostering exceptional 'esprit de corps' within units.

Aryan racial theories (borrowed from Theosophy) led to such maltreatment of Slavs that Hitler largely failed to raise the Ukraine against the Soviets.

In practise the authoritarian Nazi regime fell because of poor intelligence and imagination. The people at the top could not conceive that their adversaries had cracked the Enigma codes. They deluded themselves about their so-called super-weapons and rushed ultra-heavy tanks, jet aircraft, rocket aircraft, unguided cruise missiles, and unguided liquid fuel rockets into production in desperation. None of these 'super-weapons' proved effective because they remained under-developed and ridiculously expensive. The allies could have made such weapons but decided to finish the war with the proven equipment that worked. The Nazi's disdain for 'Jewish Science' severely undermined their efforts to make atomic bombs.

The Nazis used and misused esoteric ideas and in the end they failed horribly.

82) Wikipedia mentions a run-in you had with something called "ice magic(?)". Is that sort of similar to Helena Blavatsky and what not?

The magical order of the IOT grew into an international order when a German man Ralph and I decided to expand upon the seminars that he had organised for me in the German speaking world. He and I became the effective heads of the order. It all went very well for about five years. The order proved very productive, innovative, and creative, and a lot of fun.

Then Ralph fell under the spell of someone called Helmut whom I never met. I think I can justifiably say that because after the civil war that followed, Ralph dissociated himself from Helmut after failing to carve out a role as his chief lieutenant.

I have no clear idea of what Helmut's 'Elder Historical Tradition' or 'Ice Magic' consisted of. It seemed to embody elements of eastern martial arts, guru-yoga, Nordic apocalypticism, survivalism, standing in icy conditions, and obedience, devotion, and economic service to Helmut. It looked like a nasty manipulative cult. Helmut declined to face me after I challenged Ralph to invite him to.

Ralph began trying to take people out of the IOT and the FS where he also had a top role, and civil war ensued.

As with most civil wars, both sides committed atrocities (at least procedural atrocities) and many different perspectives on the conflict exist. Conflicts of loyalty and personality became very polarised. People flew repeatedly across the Atlantic as the loyalty and leadership of the American Temples came into question. The aether seemed thick with magical battle and propaganda and the conflict lasted for about two years. It all got as messy and complicated as a civil war can.

Nevertheless, the victors write the history. Ralph became excluded from the order. The order survived but in diminished form. Ralph made no further innovative contribution to modern magical thought or culture. He penned a small uninformative and incomprehensible book on ice magic and then fell back on recycling dull historical material on traditional magic. I retired after a somewhat pyrrhic victory to devote myself to a growing family and business, and to esoteric research.

83) Helmut, what a perfect name for a faceless villain. What would you have tried to talk about with Helmut had he met with you?

I had no intention of debating anything with Helmut. I would have disrespected and insulted him and tried to provoke a real fight. I had seen how most of these western exponents of Chinese 'internal martial arts' work. They create an elaborate charade designed to elicit obedience from their students and to make their students believe that the master has amazing powers. People had told me enough about Helmut to confirm that. They also said he looked ridiculously unfit and carried a great deal of flab. He wasn't teaching fighting; he merely taught the game of ritualised play fighting. I would probably have punched him in the nuts and then gone for his eyes. In a real fight there are no rules.

84) Did you hear anything else to paint a picture of Helmut? Leather trench coat? Hitler youth hair cut? Albino maybe?

Nothing so exotic. I have seen pictures of him, just a big flabby guy with glasses. In my generation kids like that often got pushed around at school and when they grew to full adult size, they sometimes developed a taste for control and dominance. Ah, the banality of evil.

85) Did he do some fucked up shit to some of your homies or he

was just super shitty in general? I don't suppose you make a habit of going for people's eyes?

A number of the members of the order that Ralph took to meet Helmut came to regard Helmut as a really nasty manipulative piece of work.

86) I don't suppose you make a habit of going for people's eyes?

I usually prefer to fight by grappling. I prefer to win by surprise maneuver rather than attrition. Fighting like a woman can create a useful surprise.

87) What do you think the soundtrack to that fight would have been?

Carmina Burana - Fortuna Imperatrix Mundi, or if not, perhaps Street Fighting Man by the Rolling Stones.

88) So, this would have been some time in the late 90s when you were trying to fight a German cult leader named Helmut?

Yes, the nineties. I didn't care what Helmut did with his own cult, but I did care about what Ralph wanted to do to in his name, to the non-cultish organization that we had created.

89) Thoughts on cults?

I once stood up and said to a member of the IOT who suggested that we form 'A Cult of Baphomet', that 'This is supposed to be the cult of all people who don't want to be in a cult' and got a round of applause for it. Cults depend on human suggestibility and gullibility and the need to believe in something or someone, and to belong, and often on the need to have something to hate and fight against. Cultism runs on a spectrum from harmless social and special interest groups to murderous religious and political ideologies. The fanaticism or cynicism of cult leaders often remains unknown, some seem to have enough self-belief for both.

90) Do you think it is fair to call Blavatsky racist?

In imperial and colonial times just about everyone assumed that the possession of ships, armaments, technology, and the political and religious culture that backed up their use, proved the general superiority of one group of people over another.

Religion however only exists because of doubt. Nobody invests faith in the absence of doubt.

Missionary work obviously accompanied the imperial expansions, but some found the religious and mystical traditions of the orient intriguing and wondered if they offered anything worth pillaging. Blavatsky decided to try marketing a bowdlerised Indian spirituality to the west. To make this more acceptable she adopted an idea from the linguists who had a theory about Indo-European or Indo-Aryan languages, which sort of made them our ancestors.

91) What do you think are some of the most interesting periods or movements magically speaking in the Christian church?

Christianity has always expected miracles from its saints and relics and sacraments. The common people have always expected a bit of 'magic' from Christian priests and asked for blessings of every imaginable kind and the priests have tried to oblige. Priests probably authored all the grimoires. I find the history of the Church and the Papacy, the Reformation and the Counter-Reformation, the Borgia Popes, the Inquisition, the Protestant breakaway, and modern variants of Christianity all make a fascinating tale of human suggestibility, gullibility, stupidity, greed, and politics.

92) What music were you listening to around the time chaos magic was created?

Art-Rock, in small doses. I generally dislike the trivialisation and the modern omnipresence of popular music. I would rather prefer to give a piece of Art-Rock or Classical Music my full attention for an hour or so just twice each week.

93) Were the situationists onto anything that interests you, like the idea of trying to re mystify everyday landscapes?

We pragmatic British despise French intellectual bullshit. Situationism just consisted of Hippy Marxism. I consider myself an Anti-Marxist Hippy who does not like dope. Psychogeography looks like fun for those who need it, but for me the natural and the human world seem deeply intriguing and mysterious enough just to my ordinary perception.

94) Does a familiarity with getting oneself into altered states give a person protective insight into cult manipulation?

Perhaps it can. Once you understand the mechanisms that implant belief and cause 'mystical states' you can become more discriminating in your choice of beliefs.

95) Have you seen the chart cult expert Steven Hassan uses to measure how coercive a relationship is?

I have not read his work, but I can see where he is coming from with his BITE Model (Behavior, Information, Thought, and Emotional control).

Perhaps we should consider how much BITE our own cultures and political states exert on us!

96) Yeah, the bite model I think is what I'm talking about, sorry to

say his book is absolutely unreadable, but how applicable his model seemed to politics really interested me.

The power relationships between capital and labor, between men and women, between adults and children, between minorities and majorities, change with such bewildering speed today that we need a daily weather report on prevailing moral trends. Or perhaps not, the people behind commerce and the electronic media have already made a hugely successful power grab, they BITE us in the ass to an unprecedented extent now. We have become their product.

97) Have you studied or trained in any fighting styles?

A bit of Judo and Karate. A bit of involuntary anti-skinhead work on the streets. We used to have this ghastly youth subculture of skinheads. Shaven headed nihilistic working-class youths in big boots who seemed to hate everyone and everything, they particularly liked attacking Hippies.

98) Before we leave conspiracy theory for good, as someone with enough knowledge of physics to construct a cosmological model, do you think the planes that hit the towers on 9/11 could have caused the towers to completely collapse in on themselves straight down like they did?

Any fool can design a building that will not fall down. It takes an architect to design a building that will only just not fall down. Plainly the architects had not factored in the impact of a big jet with a full fuel load under full throttle. Concrete and stone have awesome compression resistance, but steel framed buildings do not. Masonry tumbles, steel crumples.

99) Do you think prince Andrew or whichever one diddled those kids?

Prince Andrew seems pompous, haughty, charmless, socially inept, and rather dimwitted. I think he got sucked in way out of his depth by sharks who flattered and exploited him. I suspect he deluded himself that Epstein's exploited girls liked him, and he probably failed to realize or ask their ages. Thankfully, Her Majesty has two useful children, two out of four, Edward seems a hopeless case.

100) Do you ever trip out on having a royal family? I often wonder what it would do to one's world view to grow up with a king and or queen.

Our modern constitutional monarchy represents the most highly evolved and sophisticated political system on the planet. The monarchy has no real power except to deny power to others. Apart from that it merely provides entertainment and a sense of continuity. Ultimately it protects us from elective dictatorship, revolutions, or a Prime Minister going mad. The military swears loyalty to the crown. The government pays the crown, but the crown can in principle dismiss the government. It works as a marvelous system of checks and balances, even though it may look bizarre and logically indefensible. Long live the Queen.

101) What political philosophy do you most identify with?

I think the classical Greeks worked out that the ideal system consists of a balance of Monarchy, Aristocracy, and Democracy. Too much democracy leads to mob-rule, too much aristocracy leads to oligarchy, too much monarchy leads to despotism.

In practice the aristocracy has always been something you can become promoted to or demoted from.

As Churchill observed – 'the only thing worse than Democracy is every other system we ever tried'.

Nevertheless, within democracy, I favor liberty over equality most of the time.

Interview 3

102) So, you don't think there was anything fishy about those towers coming down?

No, we can build steel framed skyscrapers close together in the knowledge that if they fail, they will crumple rather than tumble and knock over others like dominoes. The fanatic Jihad cultists who attacked the towers cleverly exploited weaknesses in a situation that nobody else spotted. The Jihadists briefly outsmarted us. They may do so again in some other way. A free society with finite resources cannot anticipate or close every possible loophole.

We cannot make any building impregnable. The Great Pyramid is the strongest building humanity ever made and also in real terms the most expensive.

Conspiracy theories all seem to depend on the following dubious principle:

'Extraordinary Events Require Extraordinary Explanations.'

I remain a great fan of Occam's Razor, we should always look for the simplest answers.

A great deal of occult thinking and theory uses unnecessarily complicated explanations. In my books I have always sought to explain magic using the minimum possible amount of weird metaphysics.

Religion of course uses ridiculously complicated explanations for events. Once you put gods and spirits in charge of events you simply open a series of ever larger cans of worms and never get a useful answer.

Science also sometimes gets it wrong. Faced with galaxies that do not rotate according to theory do you modify the theory, or do you fill up the universe with a theoretical dark matter that must have absurdly complicated properties to account for the discrepancy?

103) You seem pretty suspicious of power and abstract capital to have such a positive view of your government, what am I not seeing?

I have no problem with power and capital so long as they circulate, it's when they get stuck in the same hands or cliques for too long that problems arise. At present we have a relatively new government with bold ideas for many improvements and I remain cautiously optimistic.

104) I don't know man, here I can buy guns and weed in the same day and a few states over they have prostitutes, seems like we've got it all figured out.

Well, those three freedoms can have an impact on other people's freedoms. In the UK we currently don't have complete freedom for those three activities because we consider that they will negatively impact other people's freedoms, like freedom from urban gunfire, freedom from the company of psychotic people, freedom from coerced relationships – most prostitutes would do almost anything else for the same money. I have met a few socially over the years – they hated the work.

105) So, what are the downsides to like Denmark or Sweden, huge social safety net sort of countries?

I have never been to either country, we get almost no news from them, they do not exactly function as centers of vibrant contemporary culture or places of interest except for the scenery or perhaps the history. The descendants of the Vikings seem to have opted for a quieter lifestyle for the present.

106) Thoughts on social safety nets?

Two million vicious years of the survival of the fittest has made the human race of today.

However, things now change so fast we cannot guess what characteristics will prove to be the fittest, even in the near future.

So, I suppose we may as well keep even the most seemingly useless and dysfunctional alive just in case. We have so much slack in our economies today that it costs us little in real terms.

107) Thoughts on Brexit?

I regard our escape from the EU as historically important as our rejection of Popery, and our defeats of the Spanish Armada, Napoleon, The Kaiser, and Hitler. We have reclaimed our Democracy from the jaws of a corrupt, incompetent, bureaucratic Crony-Synarchy.

The European parliament remains a powerless sham-parliament. The real power in the EU has always lain with unelected and un-sackable and largely faceless bureaucrats, and with the Germans because of the strength of their economy.

It may hurt a bit economically for a while, but any nation that trades its freedom for economic gain will end up losing both.

108) Is the current utility of the royal family just sort of how things ended up, or is there something to the whole context of a royal family that fits them for their role in the checks and balances?

Genetics count for nothing, heredity counts for little, whenever we have run out of monarchs, we have adopted one from abroad. It is the institution and the training for it that matters. No matter how good your starting material, sooner or later a meathead will be born. We have killed off several during our history and several times brought in replacements from abroad.

109) Where does your government need to improve?

The Conservatives need to take their pledge to become 'one nation conservatives' seriously and address the conditions in the old decayed industrial heartlands. The Labour party needs to concern itself less with 'woke' causes and phony intellectual Champagne Socialists. The Liberal Democrats should do everyone a favor and cease to exist. Nobody knows what they stand for, and neither do they, they merely soak up the 'neither of the above' vote.

Labour tends to dominate the urban areas; the conservatives dominate the suburban and rural areas.

110) What areas do local government fail there?

It seems to conduct most of its business behind closed doors. The media take little interest in local government. In local elections people tend to vote on the basis of their perceptions of national politics and parties – very few people even know the names of their local councilors. Local councils employ quite a lot of people, but they very rarely sack anyone for incompetence, the jobs pay poorly but they have generous pensions. Property developers with lots of money always seem to get what they want.

All of the above combine to create massive inefficiency and lack of accountability.

111) Is everyday alienation and crushing boredom a personal or social issue?

Not for me. I have hundreds of things I want to accomplish with the rest of my time. I would appreciate a little boredom to give me time to think them all through.

Boredom arises from doing what others tell you to do or doing what others say you should do. Those who profess to experience boredom and alienation should explore their own imaginations rather than trying to live up to the expectations of others.

112) Alienation and crushing boredom are problems for other people, do you see any sort of social component to that or are those people just boring and dumb?

We live in a consumer society. People get alienated from and bored with consumption.

I prefer to live as a producer. I created my own job and business, my own 'religion', my own science, my own form of magic. I make my own board games and household ornaments and small pieces of furniture. I make my own magical tools and instruments and jewelry. I value anything I made (no matter how badly) far above anything I have merely bought. I'm not really into music but if I was, I would learn to play and compose it. I have no interest in watching anyone play sport, but I do a lot of outdoor activities. I spend more time writing than reading. I consume very little prepackaged entertainment; I make my own.

I can think of nothing that would make me feel more bored and alienated than living as a passively consuming punter.

113) You know one of the things I like about some conspiracies is when they get so absurdly complex as to resemble some sort of scientific formula.

Beware of overly complicated scientific formulae, they usually just consist of recipes for describing phenomena where the scientists have not really penetrated to the underlying simplicity yet. Before Newton worked out gravity, we used to have Ptolemaic Epicycles of huge complexity to describe the solar system. Newton's equations fit on a single page. Currently the official description of fundamental particles – quantum field theory – looks like a ghastly mathematically intractable mess. Insiders admit as much.

'Nature is subtle, but she is not malicious' – Einstein.

114) Well, judging a Denmark or Sweden by the happiness of their citizens suggests they are doing something right, sometimes no news is good news. What are the important metrics by which to measure a government and how does yours stand out on those metrics?

All sorts of geographical, social, economic, technical, epidemiological factors and the rise and fall of surrounding nations and empires contributed to the high days when Scandinavian Vikings marauded and traded from Newfoundland to Kiev and Byzantium, and the latter days in which Scandinavia became a wealthy dull egalitarian cultural backwater on the fringes of Europe with low levels of social friction.

The British underclass would probably have a better life if they decamped to Scandinavia, but our entrepreneurs would find it stifling. So would I.

I get a lot of magical correspondence from many countries around the world, but little from Scandinavia.

Britain continues its relative decline on the world stage since its imperial heyday, some here want to go down the Scandinavian route, but we probably have too high a population density for that now.

115) Is there an operational equivalent to the role that the royal family plays in other governments?

Yes indeed. The Republic of Ireland has a President with a role almost indistinguishable from that of our constitutional monarch. The Irish President protects the constitution and can dismiss the government and has a ceremonial function, but no other real power. Guess where they got this idea from.

116) What sort of training is important for that role?

You must learn to endure excruciating boredom, to remain outwardly unemotional no matter what, to make conversation without saying anything substantial, to open ghastly architecture whilst faintly praising it, to receive and politely entertain some of the world's worst despots, to never express an opinion or seem to take sides. To never have any privacy or freedom outside of your own palace or castle. To have to endure being famous only for being famous, and to live a life of solemn pretense. I would not take the job.

Of course, if you become a middle eastern Monarch or a Third World Dictator you can still get away with despotism; but watch out for assassins and coups.

117) Thoughts on direct democracy?

In the classical Greek ideal of government by a mixture of monarchy, aristocracy and democracy, part of the role of monarchy and aristocracy is to prevent democracy turning into mob-rule. Successful democracy depends on majorities respecting minorities. Too much instantly available democracy has its downsides, it can lead to endless conflict between vociferous minorities and the dictatorship of the majority.

118) Thoughts on communism?

Apologists for communism often claim that it has never been properly tried. As all experiments have failed badly so far, I have extremely low expectations of it. People do not seem born equal so communism has to enforce equality and the enforcers end up being a lot more equal than others.

119) Thoughts on anarchism?

It depends on what you mean by anarchy. It often comes down to people kicking in your door and saying, 'there is no government, there are no police, we can do what we like'. If government does not have monopoly of force, warlords and thugs will establish it locally. Government is a protection racket, but the alternative is far worse.

120) When and where were you born?

In January 1953 on the South Coast of England, not far from the sea.

121) What was the reality tunnel you were born into? Class, religion, world view, etc. What did your parents do? Siblings?

The reality tunnel of working-class life in the postwar period may seem rather dreary from today's point of view. We had open coal fires for heating, we woke to windows frosted over every winter morning, we had two television channels in black and white, ate dull utilitarian food, and took seaside holidays in Britain's lousy weather. However, we did have a lot of freedoms that we do not have now. Kids could wander off and play and explore unwatched. In the boy scouts we all carried huge sheath knives and went off into the woods with axes and matches, cut down trees and cooked and camped with almost no adult supervision.

My father had an exciting war aboard on a small aircraft carrier that went on the arctic convoys and to India and Australia. Then he spent the rest of his life working as a railway clerk and never really left the small town he grew up in. He seemed to me to lack imagination and ambition, he did not look after himself and drank himself to death by age 66. He always seemed amiable enough except perhaps towards the end, but I didn't see much of that as I'd gone to college and then settled in another part of the country. It was tough for my mother and my two much younger siblings. I barely grew up with my two younger siblings, they appeared when I was 14 and 18.

My mother made more of her education than my father, although she didn't take the opportunity of university. She pushed all four of us (I have a sister two years younger) to get to university and to move away from the small town. My mother had owned and ridden a motorcycle in the days when very few women did and set up a hairdressing business whilst bringing up her children.

My parents had an upper working-class life, but they had both come from families with lower middle-class lives. Both my sets of grandparents were pub landlords, pubs featured heavily in my youth. Both of my grandfathers seem to have appeared out of nowhere, we cannot find any ancestral records. My father's father was a big alpha male, an accomplished boxer with his own revolver and considerable business acumen He appeared out of the chaos of WW1, got my granny pregnant and moved into her mother's pub with her for the rest of his life.

My family all notionally subscribed to the Church of England. They never went to church except for weddings and funerals. Nobody ever prayed or mentioned religious ideas.

There is an old joke that 'The Church of England exists for the benefit of its non-members'.

At school and at boy scouts some of the adults tried to get the boys seriously interested in it but nearly all of us thought it uninspiring and irrelevant and silly, and we just mumbled the prayers and hymns half-heartedly.

122) Early impressions of School?

I can barely remember anything from my early education. I went to a boy's grammar school. If you passed a test at age eleven you could go to these selective state schools for the top ten percent. We had a religious education master there who tried to tell us that scripture and the god theory were true. I misbehaved and argued furiously with him. On several occasions he called in the headmaster who caned me. I considered myself caned for blasphemy.

The grammar school masters were a strange bunch, they were all university men who wanted an easy life on moderate pay, short hours, and long holidays. Some were extraordinarily eccentric, some liked boys too much, some were quite brilliant, some were hopeless at teaching.

123) Favorite topics?

I loved some subjects, particularly science, and refused to learn others. Chemistry appealed greatly. I established a laboratory at home, doing the sort of experiments that would today attract the attention of the bomb and terror squads. I liked history, I despised French as effete, I liked woodwork. I preferred individual sports like swimming and athletics over team sports.

124) General feel of your childhood? Pretty happy stuff? Important early memories? Pretty social kid? Popular? Outcast? Nerd?

There seemed a lot of friendship politics and dominance behavior going on which often ended in physical fighting and bullying up to the age of about 16. After that the conflicts and rivalries tended to become more intellectualized. I tended to remain somewhere in the middle of the dominance hierarchies and have small friendship groups often including nerds. My special social skill in those days lay in making pyrotechnics and explosives. Extroversion does not come naturally to me; I must force myself to it.

Generally, childhood and adolescence seemed agreeable enough, but my seventeenth year proved the least fun. I had stayed on at school when many had left to get jobs and money and motorbikes and girls. I was too old for youth clubs and too young for pubs and I had too many part-time jobs to try and make some pocket money.

125) What were some of those part-time jobs?

My father's mediocre income and his substantial expenditure on drink and smoke meant I never got any pocket money after the age of about 13. I did newspaper paper delivery rounds, I did Saturday jobs in shops, eventually I graduated to an all-day Saturday job helping on a soft drink lorry delivery round, a cleaning job at one pub on Sunday mornings and an evening job cleaning the cellars at the other. All this probably negatively impacted my school homework, but it did give me a great work ethic. At university I did temporary jobs during all of the holidays, the worst in a food cold storage facility at minus thirty degrees.

126) How did the 60s youth culture pop up in your life? What was your impression of hippies and drugs and all that?

In the UK 'the sixties' seemed to run from about 1968 to 1980 depending on social class. I was 15 in 1968 and something peculiar started to happen, for the first time a youth subculture developed that rejected just about everything their parent's generation stood for.

Previous subcultures such as teddy boys, mods and rockers, and skinheads had largely bought into existing cultural memes but just exaggerated them with teenage aggression. Beatniks were rather alternative-cultural, but they were far and few between.

Suddenly towards the end of the sixties a huge wave of post - beatnik ideas became popular, people grew their hair long, practiced free love, smoked dope, took LSD, professed not to care about money or careers, explored esoteric and eastern philosophies, and wore kaftans, flared jeans, and flowers in their hair, or went naked. We read strange books, preached Peace and Love, and listened to strange trippy music.

In those days, the economy had plenty of slack in it, you could easily get a cheap place to live and a casual job if you needed it. The contraceptive pill removed one of the major downsides of free love. To generalize - the first British hippies were mostly well-heeled private school dropouts and it gradually permeated down the social scale taking in middle class college kids and the young of the upper working classes like me. The Punks who came towards the end of the hippy era, seemed to represent the young of the working classes, they put a somewhat more angry and aggressive spin on the Hippy's rejection of conventional values. By then the economic situation had become tougher.

Living as a Hippy was always a matter of degree. For some it meant little more than a passing fashion statement. Many weekend hippies had straight jobs. Some went the whole way, lived in squats or teepees, did endless dope and LSD, or went overland to India. The movement had a long-term cultural impact, most of what we call New-Age culture derives from the hippies.

Personally, I didn't get deeply into the drugs, but I did get into natural products, esoteric and mystical ideas, and went overland to India. I still have a beard and hair which reaches my nipples, but these days I wear it in a pigtail.

127) What did you try first magic or drugs?

I tried some of the Paul Huson type magic before going to university. Once there I tried dope which I found confusing and rather nauseating. Later I tried LSD which seemed startling in its effects the first time, but rather wearying on a few subsequent occasions. I found psylocibin mushrooms much the same some years later. From an early age I have always found drink enjoyable, and mentally invigorating in moderation, perhaps as a result of growing up in a pub rich environment.

128) Were there philosophy or history books you were into before magic?

The film of Jason and the Argonauts made a big early impression on me. I really liked the idea in that film that the Olympians depended on our belief in them for their continued existence. Did this mean we could make gods? The myths and legends and the gods and goddesses of ancient Egypt, Greece, and the Norse peoples fascinated me from my teens. I love ideas but I rarely read raw philosophy as such.

129) What were the first books on magic you read? How did you discover those?

I used to get books out of the public library before I went up to university in London. Paul Huson's Mastering Witchcraft came as a revelation. Here was a book that said that you could actually do this sort of thing yourself! At university, the London bookshops had a wealth of source material that you don't see on the high street today. I started with Eliphas Levi and then graduated to the Grimoires, Aleister Crowley, the Golden Dawn material, Kenneth Grant, and Austin Spare. Carlos Castaneda intrigued me for a while, but doubts grew as the books progressed. Robert Anton Wilson's books made a considerable impression. In the four years I spent in London after university I went and met a lot of the people on the magic scene, Lionel Snell, Zachary Cox, Amado 'Crowley', Maxine Sanders, Stephen Skinner, and a host of lesser-known figures. I also got a few friends together for rituals and experiments.

130) So, you go abroad after college, then return to England? This is when you start to create chaos magic? What did that evolution look like? How long did it exist before the first books started to come together?

I started work on magic whilst at university and continued for four years afterwards in London. During this period, I started contributing to Ray Sherwin's New Equinox magazine and writing down my ideas for myself. I met up with Ray who lived in Yorkshire and we did some magical work together and a lot of discussion of magical ideas. Before setting off overland to India I gave Ray the manuscript for Liber Null which he said he would publish. By the time I got to Australia about a year later I received a single copy of the first edition of Liber Null, in low-tech binding with a white cover. In Australia I set up a chaos temple with a handful of locals which lasted for the nine months of my stay.

My wife and I returned to the UK via several more months in India in the Himalayas where I wrote another draft of Liber Null.

We went to live in Ray's village on our return, and Ray and I published the red cover version of Liber Null. We assembled 400 copies on Ray's kitchen table, collating, stapling, and gluing together the printed pages and covers. We sold almost all of them to The Sorcerer's Apprentice occult bookshop in Leeds. Ray never gave me a satisfactory account of how the first white edition went, he claimed to have made about a hundred and only covered costs. I have only ever seen three copies, one on sale a few years ago for more than a hundred times its cover price.

I wrote Psychonaut in Yorkshire, and The Sorcerer's Apprentice printed off some copies and sold them. I never gave them the copyright, we merely had a gentleman's agreement that he would give me some (unspecified) royalties if they sold well, I told him that my aim was to seek proper commercial publication eventually if it proved popular.

After two years in Yorkshire and quite a bit of work with a temple we set up there, my wife and I went back to Nepal and the Indian Himalayas for another year and then finally returned to live in Bristol.

I then sent copies of Liber Null and Psychonaut to Weisers, and on a recommendation I had secured from Israel Regardie via Gerald Suster, they decided to publish the two works in one volume. The Sorcerer's Apprentice tried to object to this but eventually Weisers realized they had no grounds for obstructing it. TSA had once given me a small handful of cash from the shop till for 'unspecified sales of Psychonaut', they couldn't even recall how many they had printed or sold.

131) Did it feel super important at the time?

I wrote the books primarily for myself. I wanted to sort out and organize my ideas on the subject into the book that I wished I had read first, hence the title Liber Null – book zero. I also wanted other people to read it so that I could get feedback and collaborators to take the ideas forward. It did feel super important to me, for the first time I felt I had stumbled into something I really wanted to do in life that far surpassed my interest in chemistry, science teaching, and adventure travel.

132) What did the publishing process look like and what was the reception like?

Until Weisers took over the publication the process looked amateur and shambolic. After that, things moved fast. Liber Null sold well, and its publication led to the German seminars and the establishment of the IOT internationally, plus translations into other languages began.

As a result of the work and the experience and feedback gained during the following few years, I wrote Liber Kaos which Weisers also published.

The two books seem to have sold over hundred thousand copies over the following decades; I seem to have lost track of all the foreign language editions. It had all begun with a large pile of handwritten notes and manuscripts written in a squat in South London's insalubrious Deptford and then revised in a tiny little hut high in the Indian Himalayas.

Interview 4

133) What were a few of the first magical experiences you had that convinced you magic worked? Or realized just how well it work?

I found that if I hyped myself up into a state of extreme anger then quite extraordinary poltergeist effects and apports would occur. My luck seemed to improve in response to casting sigils, but most of all I found that my meditations and invocations enormously increased my inspirations and imagination.

134) What was the independent publishing scene in the UK like at the time?

I don't know much about independent book publishing at this time. Big publishers brought out a lot of the classic texts when the occult revival began but when newbies wanted to write they seemed to start on the fanzine scene. I think maybe Mandrake of Oxford may have led the way with newly written occult books independently published.

135) Was new equinox pretty unique?

Quite a few special interest magazines seemed to pop up. The Sorcerer's Apprentice had its own 'Lamp of Thoth', various bookshops, groups and orders created their own cheaply produced pamphlets and magazines and sold them in occult and esoteric bookshops or by subscription by post. Ray Sherwin started his New Equinox as a Thelemite (Crowley fan) but it began to take on more and more Chaoist material. Eventually he handed it over to Ian Read who continued it for some years as the now legendary Chaos International.

136) Who else contributed to new equinox?

I only have a couple of CI editions left in my possession and no New

Equinoxes, I tend to be profligately generous with any book or mag that I have read. I think just about everyone on the UK Off-White Magic scene who was not a die-hard Thelemite probably contributed something. Chaos Magic partly defined itself by its opposition to Crowleyanity.

137) When and where did you meet your wife? When where you married?

My university college specialized in engineering and had a low proportion of women. After exhausting the available resources, I got a tip off from the second-year students that a teacher training college across the city had a major surplus of women, so I started attending Friday night concerts there. I rarely needed to use my return ticket on the underground. I make poor choices in women, I never seemed to pick anyone that I really got on with, so eventually I chose the one who introduced herself to me. I had an odd intuition on our first evening that it would last and forty-seven years later we are still together.

We soon moved in together and travelled all around the world. We married twelve years later with a brief civil ceremony and a small house party for friends only.

We have a marriage based on a highly efficient division of labor. She decides everything on the material plane; where we shall live, what properties we shall buy, what cars she will drive (I do not drive) who we shall socialize with, where we shall holiday, when to have children, and so on. She has the mobile phone.

I devote myself to the spiritual matters, I do the magic, the cosmology, the political philosophy and voting decisions, and raise the wealth through business.

Arguments only seem to develop if either one encroaches on the other's specialisms.

We share the housework and shopping and I assist with her

extensive gardening.

138) What was Crowley's influence on the magic scene at the time?

Aleister Crowley had an enormous influence on the occult revival that began in the late sixties, he cast a vast shadow over it. The Beatles put him in a montage picture on the cover of their 1967 Sergeant Pepper's Lonely Hearts Club Band LP. The musician Jimmy Page of Led Zeppelin really took to Crowley's ideas and for a while owned Crowley's old highland retreat at Boleskine. Huge amounts of Crowley's writings became published. Crowley's old magical order the Ordo Templi Orientis became revived and published the tarot pack he had designed with Lady Frieda Harris as the artist, but which had remained unpublished.

139) What did you get from Crowley?

Crowley learned his basic neo-platonic magical ideas from MacGregor Mathers' Golden Dawn and he never bothered to develop much magical theory after that. However, he made an extraordinary difference to western magical practice by openly advocating sex and drugs and yoga as keys to achieving a magical state of mind. Sex and drugs had of course become enormously popular again amongst the postwar generation of the baby boomers who had economic freedom and the contraceptive pill, and oriental mystical procedures had begun to interest many.

One phrase from Crowley stuck with me – 'There are two ways of becoming god, (achieving magical or mystical consciousness), the upright or the averse, let the mind become as a flame or as a pool of still water'. At a stroke this seemed to explain a great deal, it explained why activities which overly excited the mind and activities which deeply quieted the mind were both used in mystical and occult traditions around the world. Studies in neurophysiology also confirmed that both states tended to create the same single pointed mental focus which I called 'Gnosis'.

140) What did you oppose?

I didn't like the religion of Crowleyanity and the whole posthumous personality cult and belief system that went with it. Crowley went to enormous lengths to make himself into a dark messiah figure. It failed to attract anything more than a small circle of acolytes in his own lifetime but from the late sixties onwards his writings and philosophy attracted a huge fanatical following. He probably had to die for this to happen. A diehard Thelemic friend of mine, Gerald Suster, once asked Gerald Yorke who had known Crowley in life, about him. Yorke said that it surprised him that so many young people had become attracted to Crowley because they wouldn't have liked him if they had met him.

Crowley was born exceptionally rich as the son of a brewery magnate. He spent his whole life spending it and living like a psychopath. As a thrill seeker he practiced wholesale bisexuality, whoring, and he took huge quantities of drugs, and embarked upon near suicidal mountaineering and rock-climbing adventures. He played around with religious and esoteric ideas from all over the world and formed two cults to get people to do his will.

Crowley's Argentum Astrum was based on the Golden Dawn but with the veiled promise of mystical sex. It seemed designed to attract wealthy young gentlemen who had to buy Crowley's expensive privately published books and treat him as their master.

Crowley formed the Ordo Templi Orientis having become impressed with the fervor of Islam. He basically invented a parody of Islam with Basques instead of Burkas, compulsory free love, an unpleasantly crazy scripture called The Book of the Law, a holy place in Boleskine towards which worshippers had to turn, and of course himself as Prophet. It proved immensely popular during the occult revival although it was always short of compliant women. Eventually the rising tide of militant Islam caused it to fall from popularity.

Crowley didn't do much results magic, he didn't need to, having been born rich. He basically practiced mysticism and created religions to get others to do his will and to imitate him, all under the maxim 'Do what Thou wilt shall be the whole of the Law'.

I specifically rejected the Thelemic idea of True Will, the belief that you were born to do just one particular (invariably Thelemic) thing. That simply sounds like the Christian idea of Original Sin turned on its head. Indeed, so much of Crowleyanity just looked like cultish middlebrow satanism to me.

141) What's the business, where did it come from, how much magic was involved in creating it?

During our second year in Yorkshire Ray Sherwin went to Egypt and we house sat for him for a year and then went off to India again. Soon after we got back Ray opened an essential oil and perfume shop in Leeds having seen such things in Egyptian bazaars. It seemed to do quite well; Aromatherapy was just kicking off in the UK. After a year or so I rented a little lockup shop and tried the same. The shop lay almost invisible up a side street and it didn't do very well until we started selling the ready bottled oils to wholefood and health food stores around the country. Then as that took off, we diversified into readymade natural products such as creams and cosmetics. The comparative failure of the retail shop gave birth to a growing mail order and wholesale business and eventually an export business as well. By then we had moved into much larger premises and bought them. Ray's venture never progressed to anything like our scale.

I don't think any promotional material ever left the premises without me casting a spell over it.

142) When did Arcanorium College start?

I think it ran for about 12 years. I started it after doing a bit of tutoring at Robert Anton Wilson's Maybelogic online Academy and based it on a similar model. I got various friends and acquaintances to give courses and charged for membership to cover the site maintenance and tutor's modest fees. It worked brilliantly productively for some years but eventually tutors ran out of ideas for new courses and the members seemed to succumb to 'internet-itis', a shortening of attention span, a relentless thirst for novelty, and an unwillingness to persist with a course of work or study. So, I closed it and now simply offer the Batchelor of Magic course by email. Many start this but few persist with it.

143) How many people attended at its high point?

Well over 50 at its height as I recall, it took up most of my evenings. I wrote two books as a result of ideas that emerged there.

144) Ok so some 300,000 years ago ochre starts being used to dye clothes and paint pictures. Agriculture, metallurgy, then about 5,500 years ago writing shows up. During that period what was early human's relationship to the phenomena we now call chaos magic do you think? Or some probable characteristics of the era? The next evolution comes with cities and empires? What do you think are some defining characteristics of that change and era?

We can only guess at the thoughts of historical pre-literate and early literate peoples, we can decipher Egyptian hieroglyphics that merely record names, battles, and dates etcetera, these seem unequivocal enough, but their more esoteric inscriptions do not. We can look at the rapidly vanishing cultures of a few groups that maintained hunter-gatherer lifestyles into the modern era, but only through the lens of modern preconceptions.

I tend to think that some sort of a natural progression can occur from animism to spiritism to paganism to monotheism to atheism, and that each of these paradigms can have its own form of magic.

Animism ascribes agency to all phenomena, everything is alive and has intent to some extent, trees, rivers, animals, thunder, lightning and rain, mountains, and the sun, moon, and stars.

Spiritism goes further and ascribes anthropomorphic mind and consciousness to many of the phenomena that animism merely ascribed aliveness to.

Paganism promotes some folk spirits to deities and it also develops deities of a more abstract nature to preside over diverse activities which develop as agriculture makes 'civilisation' possible.

Monotheism develops when paganism gets too top heavy with a bewildering array of deities and a growing sense of individualized self. People opt for a simpler paradigm in which only humans and a single deity (and perhaps a few of its minions) have mind and consciousness.

Atheism rejects the idea of deity as having no explanatory power. Some atheists go so far as to reject the concepts of mind and consciousness in any form and dismiss them as mere subjective illusions that have good evolutionary survival value.

Animistic magic relies heavily on sympathy, imitation, and contagion. Spiritist magic relies heavily on conversation. Pagan magic relies heavily on sacrifice and deal making. Monotheist magic relies heavily on invoking higher powers. Atheist magic relies heavily on operator parapsychology.

Chaos Magic borrows freely from any of these paradigms. It regards belief as a tool rather than as an end in itself.

145) Aside from the evolution through types of belief onto nonbelief there are larger historical trends that have been at work to, right?

Like the first century AD with the appearance of kabala and hermeticism, one being monotheistic and the other pagan(?)?

I would question that description.

Kabala, Hermeticism, and Gnosticism all represent 'Platonic Pagan Monotheism'.

They all have some sort of Highest Power and some sort of hierarchy of powers connecting it to humanity. These philosophies developed in response to the productive interface between Hellenic and Judaic religious and mystical cultures. In the end the monotheism more or less won but Christianity retained a lot of Neo-Platonist ideas in its sacraments and relics and in its almost pagan 'pantheon' of angels and saints

146) What is the difference between a spirit and a deity?

The amount of importance and attention devotees ascribe to it.

The gods of one religion tend to become the devils or minor spirits of the religion which replaces it.

147) So, will chaos magic one day be the defining spiritual approach of an era?

I think that for many people in westernized cultures it already functions as such even if they do not recognize it. Today we have a pick and mix approach to esoterics, borrowing freely from shamanism, tantra, yoga, and any imaginable form of supposed ancient wisdom, alternative science, and any number of pop psychologies. People speak of reinventing themselves and of the utility of various belief systems rather than of their truths.

We have an amazing number of ideas to choose from and the economic and political slack to allow experimentation. MacGregor Mathers represents to me the first chaos magician. He disguised what he was doing by playing the 'hidden chiefs' card and deploying a lot of perhaps questionable scholarship, but he created an extraordinarily eclectic mash up of just about every esoteric system then known under a thin umbrella of neo-kabala.

Something similar happened in the late Roman world. So many foreign peoples and ideas and religions had become imported or incorporated into the empire that it provided a dynamic spiritual marketplace with huge numbers of minority religions and new cults springing up, and the first real 'occult revival' began in the first and second centuries AD.

However, it did not last, and our civilization may not last, the empire crumbled, the economic and political slack disappeared, the freedom of belief soon followed, as did the dark ages.

Whilst current socio-political and economic trends continue, I would expect the chaos magic paradigm to increasingly underlie how we think about ourselves and what we choose to believe.

148) What sort of cultural effects might that have?

Science and technology can bring freedom from absolute economic hardship and political and religious freedom, plus the new information technologies have made an unprecedented number of ideas, news, and opinions available. Information technology increasingly affects even who we socialize and breed with. Identity politics, culture wars, and political polarization all become amplified by IT. The printing press did much to facilitate the Protestant Reformation and the most hardcore protestants rejected priesthoods and asserted that everyman could read the bible and become his own priest, although they backtracked on that rather quickly.

Over faced with choices, our westernized cultures have tended to fragment into a thousand shards and identity bubbles united only by economic interdependence and a legal code.

In such a scenario, the chameleon like Chaoist approach of situational belief becomes increasingly appealing. When you can see all five sides of an argument simultaneously Nothing has ultimate truth, everything remains possible.

The maintenance of a self-consistent identity or set of opinions or beliefs becomes self-defeating and sanity threatening under information overload.

Of course, a crumbling of our current civilization would probably propel the survivors into a much sterner and restrictive view of themselves and society. At present, climate, population, resource depletion, biodiversity loss, and the fragility of our IT systems suggest that it has as many fail-lethal modes as any historical civilization.

149) What sort of stuff is in Paul Huson's mastering witchcraft?

Forty years on I cannot recall many details, however the book remains in print in its umpteenth edition. I do recall it as the first book I read that said you can do this for yourself and you do it something like this. I think I remember his phrase for consecrating a wand, 'this is my magic wand, I hope it works'. I'm probably on about my twentieth self-made wand now, I tend to upgrade it every couple of years.

150) Practical application of exorcism?

I have always had more interest in 'In-sorcism' rather than Exorcism. I prefer invoking more useful 'spirits' into the pantheon of my selves or into other volunteers. I don't generally hold with trying to remove any part of ourselves, it seems wiser to acknowledge all of our existing capacities and to add better ones.

151) Have you had a reason to perform any?

The classic form of catholic christian exorcism tries to change the victim's behavior and beliefs by dominating the victim into behaving and believing differently. This involves framing the whole operation in terms of one spirit, usually the supreme spirit, dominating a lesser one. Historically, coercive physical procedures and torture were often employed against heresy, madness, and mental illness. Today the church generally shies away from hardcore psychiatric work and only offers exorcism as a form of behavior change therapy to those that actively request it.

I did once try to help someone with a head full of unpleasantly crazy ideas and paranoia. Reasoning with him proved ineffectual so I personified his madness back at him in wildly amplified form. It worked astonishingly well, he suddenly became completely rational and sensible and tried to talk me out of my crazy ravings. I quickly reverted to normal, and we spent a while discussing how he was going to get his life back together. Unfortunately, within twenty minutes he drifted back into psychosis and left to wander the streets again.

Some years later I spoke with a Swiss psychiatrist at one of my seminars about the case. He said that a shock can often flip people out of a psychosis for a while and that the result didn't surprise him. Psychiatry itself has a long history of rather shocking 'treatments', as does exorcism. Westernized societies have generally despaired of quick fixes and opted for long term chemical stupefaction.

152) Are there any historical examples of crime magic, like that practiced by Mexican drug cartels, that you know of?

Criminal gang rituals typically serve to instill obedience, loyalty, fearlessness, and often to incriminate participants. Ditto for almost any army.

Believing you have a god on your side or that the local spirits want you to throw off your oppressors, or that demons will make you successful, may not confer actual immunity to blades or bullets, but it can make all the difference in a headlong charge or a tight corner.

153) Anything interesting about criminal magic?

Confidence tricksters and magicians share many techniques. Invoking the required mindset and persona in oneself before leading the audience's perceptions and beliefs in the desired direction remain the stock in trade of both professions.

Imaginary phenomena can have very real effects. Crowley once allegedly said to Dion Fortune 'Magic is something we do to ourselves'. Yet for Crowley it so often seemed a prelude to doing it to someone else.

154) What was going on around the time of Giordano Bruno that caused him to almost be taken seriously by the Vatican? Was he into anything that filtered down to you? What about John Dee, what cultural influences allowed him his status? Was he into anything that filtered down?

I think that both Bruno and Dee asked the very reasonable question that occurs to most smart schoolkids: –

Can an Omniscient Deity answer our questions about the world and the universe?

The Vatican took extreme exception to anyone putting forward any sort of speculation or scientific opinion that contradicted the view of the world and the universe it had distilled from scripture. The Vatican took Bruno's challenge to its authority very seriously and killed him for it. It took centuries for the church of Rome to acknowledge that Copernicus and Galileo got it right and that Bruno might have had had a point about other habitable worlds.

Bruno had some interesting, imaginative, and heretical ideas, yet we do not know how he came to them, or precisely what the sorcery of which the church accused him consisted of. Nevertheless, he seems to have asserted a position of pantheism or pan-deism and derived his inspirations from that. The church took a very dim view of unauthorized mysticism and burnt him at the stake. Posthumously he has become something of a hero of free thought.

The catholic church always wanted to keep the universe small and earth-centric because it had a rather small god with such an intense interest in the minutiae of human behavior that it seemed unimaginable that it practiced such vigilance over billions of worlds and species.

Dee had it easier in Protestant England and east-central Europe. He seems one of the most learned men of his times and perhaps also one of the most credulous. Some of his voluminous writings have survived but much became lost. His contribution to the success of Elizabethan England remains a subject of partisan debate. Some credit him with providing the mathematics, instrumentation, and cartography that made the voyages of exploration and conquest possible. Some credit him with organizing Elizabeth's spy network, he certainly worked as her close advisor and court magician and astrologer for some years.

I have had a look at both the published Dee material and some of the unpublished material that one of Ray Sherwin's correspondents accessed from venerable old libraries. All in all, it consists of a mixture of high-flown complex symbology and what looks like some very silly parlor game spiritualism.

Dee employed other people to scry the spirit realm for him and many think he remained insufficiently critical and over credulous about the results.

The destruction of his extensive library and papers at Mortlake during his journeys in Poland, and his fall from favor into poverty and obscurity under King James mean that we may never know his full significance or all the sources of his knowledge and skills.

I have used his Enochian Language and alphabet to compose incantations and spells. It certainly sounds appropriate and seems to do the trick, but eventually I switched to self-devised Ouranian-Barbaric. Both languages have no history but serve to 'ensigilise' thoughts in sonorous words that have no meaning to the conscious mind.

155) Austin Osman Spare, what his influence on the occult tradition and what was his influence on you and chaos magic?

In his own lifetime Spare made little impact on the occult tradition but towards the end of it he met up with Kenneth Grant who featured him in several of the many books he went on to write in the 1970s. These books led to an explosion of interest in Spare and he posthumously became recognized as a major figure in the development of occult thought and practice and his artwork now attracts wealthy esoteric collectors.

Spare had briefly joined Crowley's Argentum Astrum, but he came to dislike ceremonial magic and fell out with Crowley.

Disdaining the trappings of formal ritual Spare developed a deceptively simple method of casting spells using what he called 'sigils'. These consist of desires written out and then reduced to apparently meaningless glyphs upon which the magician concentrates profoundly to activate the desire in the subconscious mind. Although he reportedly referred to Freud and Jung as Fraud and Junk, he placed great store on the subconscious mind and used automatic drawing as another method of liberating its powers.

It seemed to me that at a stroke Spare had identified the basic magical tactic behind all spells and incantations. The magician concentrates on something symbolic of, or analogous to, the desire rather than directly upon the actual desire. This sleight of mind prevents the conscious mind from messing up the desire with extraneous or discursive thinking and it brings the subconscious into parapsychological action.

I simply added the idea of creating concentration by a whole range of excitatory or inhibitory techniques of 'gnosis' and generalized the technique to include other methods of creating abstracted representations of desire.

Spare also had a complex and highly idiosyncratic metaphysical/psychological philosophy which he wrote about at length in rather peculiar mystical language. Occult scholars debate its meaning to this day.

Interestingly I found a first edition of Spare's Book of Pleasure in the state library of New South Wales in Sydney Australia back around 1980. It had once belonged to Aleister Crowley who had fixed his metal foil Baphomet seal inside and written an inscription signed with the elevenfold cross, the inscription read:

'Imitated from the works of Aleister Crowley, Kwang Zse and other adepts. Their words and thoughts are misrepresented and distorted. Spare was at one time a student of Fra P but was kept back by him on account of his tendency to Black Magic. This tendency is seen in its development in this book. Critics will note the provenance of the meaning of words Eg obsessions incarnating. One also finds sentences without verbs & pure nonsense like 'Neither Neither' stolen from J M Barries 'Never Never'.

Fra P means Frater Pedurabo, Crowley's personal motto. By Black Magic Crowley probably means results orientated magic for worldly desires, something Crowley usually disdained to do, and it probably also reflects Spare's antipathy towards Crowleyan mysticism.

The library could not tell me how they came to have this book, but

I suspect that Crowley's one time muse Leila Waddell may have taken it with her on her return to her native Australia in 1923.

156) Ha-ha wait so Crowley, who reveled in being a pretty huge piece of shit, thought of himself as a 'white' magician?

Crowley seemed to consider himself holy, sometimes it seems in a rather tongue in cheek way, and sometimes in the heroic sense of reveling in every facet of himself despite social conventions that forbade the expression of many of those facets. He liked sex, ordinary and perverse, and he liked drugs and he liked religion, magic, and mysticism, and saw no reason not to mix them all together. Being born rich he had little cause to use magic to gain material advantage and he records little results magic. I think he disdained it partly because for most of his life he could buy anything he wanted, and partly because he feared it might fail, and partly because he preferred to use his powers to achieve mystical states and visions, and feelings of great self-importance.

Spare on the other hand was always in need of money and the basics of life and all his magical efforts seemed devoted to either that or to inspiring his art and finding a market for it.

Crowley could afford to disdain Spare for 'grubby sorcery', and I suspect Spare refused to give Crowley enough of the worship and obedience he so deeply craved.

157) Did Spare's metaphysical/psychological philosophy have a base in other sources people could look into or is it pretty much of his invention?

Spare seems to have had a familiarity with the occult themes of his day and apparently studied Theosophy, Levi, and Cornelius Agrippa and would have become exposed to Golden Dawn style magic reflected in Crowley's Argentum Astrum. However, he seems equally if not more heavily influenced by Freudianism or at least his own interpretations of it, and the myths of Witchcraft. Plus, perhaps elements of Buddhist and Taoist thinking entered his philosophy. His obscure passages about the Kia and the 'Neither-Neither' seem to relate to the ideas of formless spirit and oriental 'Not this - Not that' meditation techniques.

158) Do you have some favorite works of his? Is his work easy to come by?

'The Book of Pleasure' remains his masterwork. It details his techniques of sigils and automatic drawing and it also rambles through his philosophy and metaphysics using highly idiosyncratic language which remains open to interpretation.

First editions now attract crazy prices, but several reprint runs have appeared over the years.

His art has become collectable and expensive with some seeing him as an early exponent of surrealism and efforts to represent the subconscious.

159) Did you only teach abroad?

I didn't do any academic teaching abroad. After finishing my chemistry & biology degree, I knew only that I didn't want to work in a laboratory or industry. My girlfriend, then in teacher training, suggested teaching and I thought why not, it's another year of student grant and then perhaps a job with short hours and long holidays.

I spent a year on a postgraduate teaching certificate. The Bursar of the college welcomed new students with a completely forgettable speech, except for the opening sentence – 'We cannot teach you how to teach'. They didn't bother. The course consisted of two bouts of teaching practice and some time in college discussing the latest arbitrary fashions and theories in educational philosophy. After the first teaching practice everyone piled into the Psychology of Education lecture option. The lecturer opened with 'I know why you are all here, this happens every year, you all want to know how to psych the kids don't you? I have to tell you that I don't know'.

Basically, if you survived the two teaching practices and wrote a couple of essays reflecting the academic's ideas back at them, you got the certificate. In front of classes I simply acted as I remembered my old schoolmasters had acted.

160) How was that experience?

After a long hot delightful summer hitch hiking around the UK, I suddenly realized I had better get a job. I put in a single rather late application backed up by spells and ended up as the only candidate for a post at an all-girls grammar school (one of the last of its kind) run by a coterie of dear old spinsters who had been there for many decades since the days when female schoolteachers were expected to remain single. I think they took one disapproving look at the young male hippy and decided they would have to make the best of it for a year as nobody else had shown up and it was nearly September.

In the event I stayed for two years. The first was very civilized, Latin hymns at assembly, tea in chintz cups in the staffroom, the pupils generally studious and well behaved. In my second year there, the place erupted into mayhem after a forced merger with a really rough non-selective school in a poorer area about a mile and a half away. The new kids ran wild, pulling the wigs off several of the grand old dames who ran the place and telling them to F-off to their faces. I had to rescue my dear old department head from a braying mob of them.

After that year, a lot of the old ladies resigned and I left to hitch hike to India with my girlfriend, having saved quite a bit of money because of living in a rent-free squat.

On our eventual return to the UK, we settled in Yorkshire for a couple of years and there I got an even more peculiar teaching post in an all-boys Roman Catholic school where extraordinary medieval practices were maintained. The Headmaster and his deputy were both priests and completely bonkers and out of touch with contemporary reality. Most of the staff were old boys of the school. To encourage them to stay on for the sixth form, smoking was encouraged in the sixth form common room. The headmaster distributed cigarettes to boys of all ages as part of his 'pastoral work'. The whole place ran on casual violence. The boys asked me if I was a real teacher because I didn't carry a tawse (a heavy leather strap for whacking the boys). Staff would cheer on fistfights in the playground, if a kid took a swing at you, you were expected to take one back. Two or three times a year on special saint's days the boys would go home early, and the staff would get blind drunk in the staffroom on crates of sherry. I was not invited as a non-Catholic. I think they did it so that they would have something to confess later. At Easter the school conducted a full-scale re-enactment of the crucifixion with mock flagellation and the most handsome sixth former dressed in a toweling nappy hoisted up on a huge cross in the main hall for the adulation of the religious staff and the sniggers of many of the boys behind their hymn books.

The staff were full of what struck me as weird superstitious beliefs and the boys often talked about occult subjects. I kept quiet about my interests but once when doing some sort of a biology module we did some tests of memory and sensory perception and the boys asked to test extra-sensory perception. I set up a quick experiment involving one boy at the front looking at a random image from a pile of science magazines and the rest trying to draw what they thought he was looking at. The results were not impressive, and the boys said, 'we knew it was all bullshit- you have a go sir'. Feeling a bit put on the spot by the failure of the experiment I sat down with them and asked the kid at the front to pick a new image. Somehow, I just intuitively knew what he had picked and sketched it correctly. 'Bloody hell that's amazing we knew it was real' they said, and then, 'no it's a trick isn't it, how did you do it?' Saved by the bell I left them guessing. I can usually only do this sort of thing under pressure and certainly not on a repeatable basis.

On another occasion I managed to persuade a rowdy class that I was dissecting a human heart for their education. I explained that the city mortician was a close friend and had agreed to let me use the heart of a recently deceased elderly man so long as we treated it with the utmost respect and returned it for christian burial. It certainly got their undivided attention. In fact, I had simply purchased an ox heart from a butcher's shop and let it get a bit over-ripe. One kid threw up into a wastepaper basket, another threw up out of a window above the staff room, but we all had a good laugh about it when I disabused them at the end. After that I found it easier to get them to pay attention.

On returning from India a second time and settling in Bristol I worked as a supply teacher for two years. This meant going round to cover teacher absences in lousy schools where the only expectation was to keep them in the room and amuse them by any means possible. Bristol had an exceptionally high concentration of private schools and some of the worst state schools of any big city. These two things tended to reinforce each other. Local councils controlled the state schools, permanent staff were effectively un-sackable, promotion depended mainly on seniority, many headteachers adopted a purely ceremonial role, played a lot of daytime golf, and waited for retirement.

For a few terms after I started my business, I taught human biology to adults at a technical college in evening classes. The students were mostly nurses or alternative health practitioners who needed the qualification. I found it an immense pleasure to teach well behaved people who were motivated to learn, and we achieved exceptional results. Perhaps if I had gone into adult education, I might have stuck at it.

161) Any long-term insights about humans from teaching?

A motivated person can learn anything. As a schoolteacher you are expected to supply motivation to those who lack it, this becomes more difficult if you are supposed to teach material irrelevant to the lives of your pupils. I despaired of teaching science to illiterate and innumerate kids, so when I got assigned to the bottom groups nobody else wanted to teach, I brought in frying pans and taught them how to fry eggs over a Bunsen burner, and other skills like growing vegetables behind the squash courts. The one size fits all model of education simply doesn't work. It wasn't supposed to, it was designed to grade people for success or failure.

Educating the sexes separately between the ages of eleven to eighteen gives better results than co-education, particularly for girls. Between these ages the sexes mature at different rates anyway and taking teenage sexual politics out of the equation within schools simplifies their social management. Plus, single sex schools can be smaller and still offer a full curriculum. Smaller schools usually work better, they can function as proper communities, whereas large schools tend to become mere educational factories where the downsides of anonymity outweigh the economies of scale. I have worked in huge shambolic schools where few of the staff knew the majority of other staff or more than five percent of the pupils by name or face.

Schoolteachers work like sheepdogs; they play on the sheep's respect of wolves. Calling in the parents remains the ultimate sanction. If kids have no respect and fear of their parents, the job becomes difficult.

As a schoolteacher I acquired more distaste for schoolteachers than I had as a pupil. So many of them seem incapable of dropping the act when off duty. The games staff usually seemed the worst, because they had few career prospects after 45 unless they secured a pastoral post, they tended to become the self-appointed moral guardians of the school's ethos and harped on interminably about 'attitude'. Most school team sports seemed to have strong undercurrents of organized bullying and dominance play.

162) Had you seen much of the world before going to India?

Not really, before university I had hardly seen anything beyond southern England. In my early college summer holidays, I mainly worked but hitch hiked to hippy Amsterdam which proved interesting and later also to the Munich beer festival which didn't, apart from a brief fling with an American lady who was equally bored with the idiotic drinking and fighting.

Then in later holidays with my girlfriend I went around a lot of the UK from Cornwall to the Orkney Islands and later also to Morocco.

We went to Morocco during a summertime Ramadan. Morocco should be a major tourist destination from Europe, it has fabulous beaches and mountains and lots of historical interest, but the locals take a low view of tourists and do all they can to exploit them in a very shortsighted way, trying to rip them off at every turn. Summertime Ramadan fasting also made them bad tempered. We got out with nothing worse than a bad feeling about the place and a dose of dysentery, but few people go there to this day. It could be bigger than Spain for tourism if the locals changed their attitudes.

163) What is a squat? How many have you resided at? Favorite one? Any life lessons from the experience?

The law in the UK has tightened up on this in recent decades but back in the day anyone who managed to get into and occupy an empty property had certain rights and it often took a long while to legally evict them. In theory they could eventually become the legal owners if they managed to remain there for a sufficient number of years. Quite a few alternative types did this, sometimes in prestigious properties with absentee foreign owners in London. Squats varied enormously from the fairly straight to the completely bohemian, from the well-organized to the filthy and shambolic, from the ideological to the artistic, to the drug addled.

The teacher training college I went to took charge of a dilapidated block of low-rise flats in Deptford to use as student accommodation for several years before its demolition date. The college half-heartedly collected rents for a few months and then just gave upon it and the block filled up with a mixture of students, ex-students, and assorted itinerants. It proved a colorful, lively, and dangerous place. In our two years there we had a couple of murders, running street battles with groups of Irish travellers who sometimes occupied the waste ground nearby and with neo-Nazi skinhead gangs. Next to the flats we had a metal and general scrap yard run by some very heavy and dodgy people who could supply anything for a price.

164) Who were some of the more colorful or brilliant characters you met when you were starting out in the UK magic scene?

I'll tell you three tales of people who showed me what not to do with magic, and one of someone who set a shining example.

I met Charly Brewster at some small occult meeting, he was a few years older than me, he had a posh private school education, and had lived a hippy lifestyle in London for a few years. When I met him, he had just got out of jail for a spree on a credit card he found in the street.

I told him the block I lived in had an empty flat and he moved in. He had a wealth of occult knowledge and we got on brilliantly.

In jail he had taken a vocational course as an electrician. He went to work each day in an expensive suit and carrying an equally expensive briefcase containing nothing but a soldering iron. He explained to me that one should always dress for the paygrade you wish to acquire. He was a large guy and had an even larger personality and the gift of the gab. whilst I was off in India, he blagged his way into a very senior technical position at Reuters.

By the time I returned from my second trip to India he had founded his own electronic company and bought a mansion, prestige cars and powerful motorbikes. He filled the mansion with high end occult paraphernalia and became a major player on the Thelemic and Chaos style magic scenes. He had huge appetites for excess and would serve guests with rare single malt scotch whiskies in half pint mugs, before rolling enormous spliffs.

On a couple of occasions, I went with him on crazy missions to contact the gods of the Celtic underworld in deep caves beneath the Welsh mountains, he took extreme risks. With some knowledge picked up at Reuters he started dealing on the Chicago stock exchange using some madcap scheme he had thought up. He described himself as a 'crisis magician' – someone who doesn't put in daily practice but only uses it in tricky situations or emergencies, and he adopted the magical identity of Fra Choronzon 999 after one of the Enochian Demons that Crowley had a run in with. A lot of people thought that a risky move.

Eventually everything imploded under a tsunami of debt. He had built an empire founded mainly upon his ability to persuade people that he should have one. He had a fantastic imagination, but it often lost contact with reality. When things were up, he was ridiculously generous.

When things went down, he borrowed persuasively and heavily from everyone. He even ended up owing a telephone company the price of a small house as a result of his transatlantic share dealings.

When everything collapsed for him, he disappeared with his long-suffering wife and remained incommunicado with all the people he had known in London. I eventually found out that he had retreated to a quiet life in a remote part of Wales and had a couple of heart attacks, the second of which killed him.

Gerald Suster had gone to one of Britain's top public schools, (in the UK a public-school means its open only to the scions of the wealthy, so they are sometimes more realistically called private schools). He spoke beautifully and had perfect manners and came over as the archetypal perfect upper middleclass English literary gentleman. He made a living teaching English in private colleges and desperately wanted to make it as a writer. Unfortunately writing and drink and Thelemic philosophy led him to an early grave. He had a bad break when a journalist from the now defunct scandal sheet The News of the World turned up incognito at a discussion he had organized for the older pupils at the college on esoteric ideas and asked him a couple of seemingly innocuous questions and got some photographs. The following Sunday the paper published a big spread about a blood drinking Satanist working at a top private school. On Monday, the headmaster called him in and said, 'I know this is all bullshit Gerald, but I have to let you go'. Friends helped him to bring a court action for libel against the newspaper which he won but the damages he received amounted to only about two years' salary and his career as a teacher was effectively finished.

He turned to writing but his increasing obsession with Thelema and Crowley meant that he never made magic his own or wrote directly about it. He wrote a book on boxing history that apparently remains world class in its field, and a book about the history of the Tarot which was moderately good, but then he started writing what he admitted were trash novels for money. He would sit for months on end at a manual typewriter fortified by cartons of full-strength untipped cigarettes and bottles of gin churning out sex and violence laden stories with some occult themes, that he imagined people would want to read. He never found much of an audience and had to churn out several books a year to make a meagre living.

He would turn up at occult conferences and at people's houses in unpleasant alter ego mode, loudly and boorishly proclaiming Thelemic philosophy whilst swigging from a gin bottle and generally behaving in an aggressive, argumentative, and utterly ungentlemanly fashion. Before opening a fresh gin bottle, he would proclaim 'Do What Thou Wilt Shall be the Whole of the Law' – as if he were saying grace before a meal.

He developed a strong identification with Earnest Hemmingway, a famously alcoholic but rather more talented author. I last saw him at a wedding reception to which friends had charitably invited him with reluctance. At age fifty he looked like an emaciated and disheveled eighty-year-old and died some months afterwards.

Amado Crowley provided me with a classic example of the path of Charlatanry. He appeared on the scene as a chubby middle-aged man proclaiming himself as a biological son of Aleister Crowley. Crowley himself had once quipped that 'any English Master is running either a boarding house or a brothel'. Amado had at least realized the potential of occult mastership to supply him with young people to pay the rent and provide him with personal services. He had a large appetite for young men and a luxury 'commune' in the home counties. He almost certainly had worked as a psychology tutor in an adult education college and had no Crowley ancestry at all. He had cobbled together a mish mash of Crowleyan ideas, imagined historical Anglo-Saxon esoterics, and odd bits of new-age occultism, with the overriding theme that He Was The Master.

It didn't fool many for very long and he was on the continual lookout for fresh young people, mainly men. Some of my friends and I met him a few times in London and he seemed a font of intriguing ideas and magical knowledge, but when I got to go and visit his 'commune' I realized within a few hours that it was all a manipulative sham and that he was a complete psychological con artist.

Lionel Snell came from a similar educational background to Gerald, but he went on to study mathematics to the highest level. There were a lot of posh young people on scene in the early days of the occult revival, Lionel however has always remained a perfect gentleman and he became the most celebrated magical philosopher of his generation, although he retained a career as a mathematics tutor and a technical writer. He often expressed profound ideas in comic terms and wrote some hilarious material on the paradoxes of satanism in the early occult zines. He brought out a book called Sex Secrets of the Black Magicians Exposed, the contents of which had nothing to do with the come-on joke title, it was an upmarket book on the psychology and metaphysics of magic. Recently he released his magnum opus, My Years of Magical Thinking, a largely philosophical work which contrasts and compares the thinking styles of magic, science, art, and religion., and examines how they interact.

Lionel does not confine himself purely to the theory and philosophy for which he has become renowned. I have had the privilege of participating in some of his exquisitely crafted rituals and workshops.

These four characters had considerable influence on my thinking and career. Teaching magic or writing about it seemed a poor way to make a living – don't give up the day job. Thelema often acts as a very self-destructive philosophy. Magic can make a big difference in business, but you need to use it with restraint and not get carried away with it.

Interview 5

165) Did you see the recent article on a new model for a possible warp drive? Thoughts?

If you mean the Alcubierre 'warp drive' I don't hold out much hope for it becoming practical in its present theoretical form, it appears to require impossible amounts of energy or some form of 'negative energy' – whatever that means. At present neither of the two main official theories of physics, general relativity, and quantum field theory, offer much of a clue as to how we could make a practical warp drive. Presently we cannot find any way to explain gravity in the same way we describe quantum effects, or vice versa. If we develop such a description, it might possibly allow us to manipulate gravity, and hence spacetime itself, in the same sort of way as we manipulate electricity and magnetic fields.

Presently we seem stuck with reaction-thrust space craft which basically acquire momentum by throwing stuff out the back. This usually means hot gas from rockets but accelerated ions can also be used to build up momentum more slowly and persistently by craft already in space. Either way it is tremendously inefficient and even on an interplanetary scale its pathetically weak for our ambitions.

166) Thoughts on private companies now playing a large role in the space race?

The initial motivation for getting into space was largely military. If you can put something into orbit and bring it back down again then you can hit anywhere on the planet with a nuke, and you can spy on anywhere and send communications from orbit. Orbital civilian communication satellites have since proved a great commercial success and private enterprise has moved in on it, but under heavy governmental scrutiny.

Orbital space also provides a great environment for astronomy and the Hubble telescope has done wonders.

167) Excited about the potential for a moon base?

The three crewed Moon landings were driven by political motives with the USA and the USSR striving to demonstrate their prowess to the world. We already knew there was nothing worth bringing back from the moon and that un-crewed machines could have accomplished the mission at a small fraction of the price.

168) Mars colony?

Sending humans to Mars in reaction thrust vessels will prove insanely expensive, and even more so if we attempt to bring them back. Mars, like the Moon, just consists of a lifeless rock where human survival would be a desperate and hideously expensive business.

169) Are generation ships of any interest to you?

Well, they make interesting sci-fi microcosms for unusual human soap operas, but I doubt that humanity will ever dispatch any. For a whole raft of technical and economic reasons they seem likely to remain forever completely impractical.

I don't think humans will ever leave earth for somewhere else unless we discover some sort of trick that allows us to somehow jump 'across' vast distances of spacetime rather than 'through' it.

I'd like to see more effort put into finding out whether the structure of reality permits this or not.

Sending machines around the solar system will bring us interesting scientific information at a reasonable cost, but rather than see money and effort expended on wasteful crewed expeditions in rocket style craft I'd prefer to see it expended on investigating the possibility of some kind of warp drive, fold-space, teleportation, or hyperjump alternative.

170) How did your adventures with Charlie to contact Celtics gods of the underworld go? What were you contacting them about?

I cannot remember what knowledge or power Charly sought. He asked me to perform the conjurations and incantations with him as the vessel for manifestation of Gwyn ap Nudd. He became possessed for a while and I kept no record of his visions and words, but after that his meteoric rise from homeless jailbird to entrepreneur with a mansion seemed to really kick off. My enduring memory of the trip remains his near suicidal insistence that we enter the cave system via the emergency flash flood escape route without using harnesses and carabiners. At one point this involved traversing a steel wire bolted to the ceiling hand over hand, across an eighty-foot-deep chasm. For me, the terror gnosis he probably wanted to induce was somewhat ameliorated by only being able to see about fifteen feet with my carbide helmet lamp.

171) What's the relationship between important ideas and humor?

The Aha! Eureka! moment that comes with a discovery often provokes elation and laughter, particularly if a simple solution suddenly arises to what looked like an impossibly complicated problem.

I think we often laugh at the previous misunderstanding.

You don't always see the funny side of an important idea if you don't see the complicated nonsense it replaces. If you want kids to see the funny side of Newton's theory, give them half an hour of Ptolemaic Epicycles first. If you want them to see the funny side of the theory of evolution, try first explaining, in all seriousness, religious theories about sex, ageing, fossils, geology, disease, parasites, and death.

For me, the realization that I could account for all occult phenomena with the simple dictum that it's all inside our heads and our own small parapsychological abilities, still makes me smile when I contemplate the ludicrously baroque spiritual explanations offered by others. Most of the iconoclastic humor in chaos magic derives from this insight.

Recently I had a Eureka moment over the realization that everything wrong with conventional cosmology derives from a coordinate singularity mistake made a century ago. Polar bears do not shrink to the size of mice near the north pole, nor does the geography go crazy there. The lines on the map produce an artificial coordinate singularity at the poles but only because we try to draw straight lines on a curved surface, do that in curved spacetime and you will end up mistaking the map for the territory and imagining a Big-Bang. When I finally realized that cosmologists have persisted with using Schwarzschild's Interior metric for a black hole in the reference frame of an Exterior observer, I fell off my chair.

172) So, I imagine the IOT had a number of official rituals like an initiation ritual? What others did you have, were they mostly based on a golden dawn framework? What did you want to do differently with your rituals than what you saw of other fraternal magical orders?

The Neophyte ritual asked the applicant to agree to the proposition that 'There are No Ultimate Truths'. We asked this to filter out closet monotheists and infiltrators with hidden agendas. Some said, 'Okay, but is that an ultimate truth in itself?' This we regarded as acceptable.

Magister Templi who organized local temples had to appoint an official Insubordinate to criticize their work and policies.

Apart from that, the grade rituals were essentially celebrations and proclamations of achievement and intent.

The order suffered from the perennial problem of a lack of differentiation between the personality and organizational abilities required to hold groups together and the esoteric prowess and creativity required to sustain the experimental and evolving order we hoped for. People were advanced through the grades, often very rapidly, in the hope that they had some of both. It didn't always work.

The order had a ritual that it opened and closed its annual general meeting with, The Mass of Chaos. This multifunctional rite included solemn openings of the circle with visualizations and chanting, wild dancing, an invocation of Baphomet, consecration of a shared sacrament which could consist of almost any liquid, a laughter banishing and a solemn closing. Apart from that rather loosely defined and adaptable ritual, few things remained constant, everything remained experimental.

173) I heard somewhere the idea that Newton's explanation of gravity is essentially an alchemical idea that he expounded on rather little. Is that true if so where did he get that idea and what influence does it have on the way scientists currently view gravity?

In Newton's time 'Natural Philosophers' made little differentiation between what we would today describe as the separate disciplines of science, natural magic, astrology, theology, and scriptural studies. A huge variety of explanations for the doings of the phenomena of the world were played around with. Alchemists developed extraordinarily complicated ideas about the peculiar behavior of materials because they tended to focus on materials that behaved in unusual ways, like strong acids, mercury, sulfur, and antimony.

Newton's remaining papers are fragmentary, often coded with indecipherable symbols, and often contain terms the meaning of which we can only guess at. Alchemy was in a legally grey area at the time, a rather dark grey, and he had to be careful what he committed to paper.

Argument continues about the influence of what we would now call esoteric ideas on his work on gravity. The idea of an invisible and seemingly immaterial action at a distance can seem more occult than mechanical, and Newton himself felt uneasy about it. Arguably we didn't get a fully mechanical explanation of gravity until Einstein described it as the geometry of curved spacetime.

Nevertheless, Newton's work on optics, mechanics, and gravity was quantitative and mathematical. Alchemy only evolved into chemistry when natural philosophers started measuring precisely how much of something interacted with something else and precisely how much of what end products resulted. Newton's alchemical work seems largely qualitative, but the quantitative example he set with his other work set the pattern for all the sciences of the future.

174) Can you recommend some good books on cosmology?

Anything I recommend today may well be regarded as outdated next year. The official standard cosmological model called Lambda-CDM cosmology is what they currently teach in colleges and write about in popular straight science books and articles, but this model only dates back to around the turn of the millennium. Lambda stands for the dark energy, and CDM stands for cold dark matter. Officially, today these two things comprise about 95% of the universe, but in such a fast-moving field they may not even officially exist a few years hence. As Landau quipped – 'Cosmologists are seldom right but never in doubt'. Twenty-year-old textbooks remain a source of embarrassment.

Outside of the self-perpetuating academic circle paid to maintain, defend, and extend the orthodox L-CDM model you can find everything from straight academics who accept some of it but question other parts of it, to mavericks, usually outside of the sacred circle and with varying degrees of expertise, who question the basic interpretations of the observations, and most of the standard theory.

Nobody can really argue with the data coming back from telescopes and satellites, but great scope exists for arguing about how to interpret it. Do those six pixels on the screen come from a galaxy just a billion light years away or from a galaxy several times that far away? How many billions of light years do we think the source has moved since it emitted that light? Do we need to ascribe more weird properties to dark matter to balance the books or can we simply change the books and modify our expectations about gravity at long range?

I do love an active field in a state of flux.

175) What is the relationship between magic and healing? Neither new age healing or western medicine seem to be hitting the mark really and then there's the placebo effect.

Western science-based medicine works well with transmissible diseases and injury repair, but it still seems to have a long way to go with some chronic degenerative diseases. I do wonder if alternative health ideas have influenced orthodox medical practice to take more notice of the importance of measures to promote wellbeing and positive outlooks.

Westernized societies promote a lot of activities that can have injurious effects on mental health. Status acquisition and hyper-individualism don't work well for a lot of people and they end up with status anxiety and alienation. Depression has become the epidemic of modernity.

When you look at esoteric systems of chakras and chi meridians it quickly becomes apparent that such systems of esoteric bodily 'energies' contradict each other in detail. However, they do all have one thing in common, they all focus attention, visualization, and intent on the body, if only by analogy. There seems a fair bit of evidence that for some conditions the effort to remain positive and think yourself better does pay off. However there seems little evidence that you can remotely think someone else better to much effect.

176) What stuck with you most about your travels in India?

My girlfriend and I stashed our few possessions with relatives and stood on the road out of London with just a rucksack each, a small tent, some travelers cheques, a piece of cardboard with India written on it, and stuck out our thumbs. We got to Istanbul for free within a week and afterwards mainly used local busses to move from city-to-city eastwards.

Having already spent a few weeks in Morocco we felt prepared to endure contaminated food and drink and exploitative attitudes towards westerners in Islamic cultures. Turkey wasn't too bad apart from the dysentery, the eastern end where ordinary holidaymakers didn't go seemed a long way from Europe culturally.

Iran had a grim heavy atmosphere, the Shah had yet to fall, and the place felt angry and on edge. I had to walk as closely behind my girlfriend as possible to deter young guys from running in and grabbing her backside.

Afghanistan barely exists as a state, then or now, it seems to consist of a default territory that nobody can really control. Herat seemed mainly composed of crumbling dusty ruins where almost nothing worked. At a dilapidated tea shop that seemed to have only two cups the toothless old proprietor rolled a golf ball sized lump of hashish resin over the table to us. We shook our heads at it. He shrugged rather sadly and then poked the whole ball down his own throat and washed it down with a swig of tea. The owner of the dilapidated bed bug ridden hovel of a 'guest house' we stopped in, invited us to a cultural event in the evening.

Along with half a dozen other travelers we went into the courtyard that had a circle of benches and an audience of about fifty locals. The women all wore full burkas, some attempted to smoke cigarettes inside them and managed to look like Daleks on fire. The men all carried weapons and the men and women sat on separate benches. At last, the entertainer entered the courtyard and started juggling a shovel. This shovel juggling act went on repetitively for a couple of hours. I wondered if he would perhaps at some point try juggling a second shovel as well, but no, he only had the one. After a while I got up to relieve myself behind a crumbling building. When I returned another traveler had wandered in and taken my seat, so I went to plant my backside on another long bench that happened to have a local burka clad woman on the other end of it about ten feet away. Instantaneously, swords came out of sheaths and antique rifles came off shoulders. Fortunately, my backside never quite touched the seat, I put my hands up and bowed and they allowed me to back away without cutting off my balls and shoving them down my throat, a common cultural practice there apparently. We left Afghanistan via the Khyber Pass without buying any of the drugs or handmade guns the local peddlers offered us at the teashop on the summit.

We passed rapidly through Pakistan; it didn't seem like a place for hippies to hang around in. At the last overnight stop there the guest house owner said that to pass into India we would need a certificate to show that we had received the anti-louse treatment which he would provide for free. Eight of us travelers, including two blonde Scandinavian ladies, filed upstairs in turn to receive a full body massage from a burly masseur. It wasn't quite sexual assault, but it was exceptionally thorough, leaving no part of our naked bodies un-oiled.

When we waved the certificates at the Indian border guards, they burst out laughing, 'You too eh? ha ha ha, no, is not required at all Sahib, welcome to India.'

India seemed like entering paradise, apart from the terrible poverty of many of the inhabitants. We spent a few days at Amritsar viewing the extraordinary Golden Temple of the Sikhs and gorging ourselves on the first beer and bananas we had since leaving Europe. Everyone in India seemed friendly and relaxed about having hordes of hippies wandering around and doing hippy things and getting into eastern mysticism. Traveling India was simplified by the fact that everyone with even half an education spoke English. India has dozens of regional languages and English functions as the de-facto lingua franca.

We were 26, a bit older than many of the travelers and we had already developed some ideas of our own about spiritual and esoteric matters. I had studied and practiced magic, and my girlfriend had studied and practiced Tai Chi, so we didn't get quite so carried away with eastern mysticism as a lot of the other young travelers.

We went first for a few months to the Tibetan settlement of Macleod Ganj in the foothills of the Himalayas. Here the Dalai Lama and a few thousand Tibetans lived in exile in the five-thousand-foot-high forested hills with the glistening peaks of the mountains as a backdrop. Monkeys cavorted in the trees around the village and stole unguarded items through windows. Leopards sometimes took pet dogs; bears remained an occasional hazard to walkers and farmers. We once encountered a King Cobra slithering across a path twenty feet ahead of us. The Tibetans had built temples and Stupas all around the hills and monks and Lamas wandered everywhere.

Quite a few westerners had taken up residence there and got into Tibetan Buddhism to various degrees. I went to a few introductory lectures, but they seemed on a very preliminary level like Sunday school for children, so I took to reading extensively in the library they had built. This contained a copy of probably every book written about Tibet in English. Tibetan Lamaism had an important social function on the harsh Tibetan plateau. It acted to limit the population and prevent land becoming subdivided endlessly between sons over the generations. Second sons went into monasteries. The Tibetan version of Buddhism became a system of astonishing complexity and it demands practices of extreme repetitiveness. All this seemed designed to keep otherwise unemployed men constantly busy with rote learning, scripture copying, and chanting throughout the endless winters. The peasantry was expected to give food to the monasteries in return for blessings and festivals.

In the Tibetan Vajrayana form of Buddhism, the historical Buddha hardly gets a mention, but it has a huge array of Bodhisattvas, somewhat like christian saints but capable of reincarnation, whom the Lamas and the Laity revere, along with a whole array of spirits and demons seemingly inherited from the fusion of Buddhism with Asiatic shamanism. The theology and philosophy can get incredibly complicated, and the iconography is spectacular. The dominant Yellow Hat sect of which the Dalai Lama is head tend to downplay the sorcery and magic and take a more scholarly, philosophical, and academic approach, but the older Red Hat sect has more of an interest in it, and the oldest sect, the Black Hats, preserve many of the old shamanic ideas and magical practices.

I found a copy of David Snellgrove's *The Nine Ways of Bon* in the library, an academic treatise on the Bonpos who became the Black Hat sect. To me it had very much the flavor and techniques of Chaos Magic about it. The Tibetan cultural office in Macleod was unable to put me in touch with any Black Hats, maybe they didn't have any there or maybe they wanted to keep quiet about it. Allegedly the guy employed to prevent rain on festival days was one of them.

After some idyllic and fascinating months in the mountains we went on to Goa via Delhi by train, not quite understanding what second class unreserved meant. It meant third class standing room only with chickens and goats underfoot, and the disconnection of the carriage after the first hundred miles or so. My girlfriend got ushered into a women-only carriage and rather than lose contact with her I spent the rest of the night clinging to the buffers between two carriages along with a changing selection of ticketless passengers. I saved one guy's life by pulling him out of the way of a fast-approaching bridge pillar that he had not seen, he just shrugged and smiled. The Indian attitude to health and safety and fatal accidents seemed very relaxed, they all expect to reincarnate anyway. When we finally pulled into Delhi we met on the platform, she barely recognized me, it was a dirty old diesel train, I looked like a coal miner.

A few days of mass nudity and the constant fug of hashish on the Goan beaches proved quite sufficient and we fell in with a group of travelers working on a project to build a boat and sail it to the Maldives Islands. We ended up with a camp in a coconut palm grove on a riverbank in the far north of Goa building two boats. Various travelers joined the project for varying amounts of time but basically me and a Danish guy built most of one boat and my girlfriend stitched us a set of sails and a group of Germans plus an Indonesian built most of the other one. We had bought two derelict river boats based on dugout logs and then replaced the rotten timbers and added masts and bowsprits and rudders, extra freeboard, and rowing benches and outriggers. Fortunately, there was a sawmill a few miles upriver and ships chandlers to service the local fishing fleet in a town a day's bus ride away. My expertise with an adze grew to the point where I could split a matchstick lengthwise with it. After four months of work, we had two thirty-foot boats, each with an auxiliary small canoe, a crew of seven, and monsoon season alarmingly close. Nobody knew much about sailing, but we decided that we would try to sail south and round the cape of India and visit Sri

Lanka rather than risk the open ocean in search of pinprick Island's hundreds of miles out.

It all went delightfully well for about two weeks; the wind blew steadily south, and our poorly maneuverable craft went with it. Then the German boat failed to steer round a small rocky offshore island and smashed itself onto it. We rescued them and put them ashore and continued sailing south for another week, then one evening the sky suddenly went black early and a tropical typhoon blew up. Before we could react, it tore the mainsail off, snapping the 30-foot bamboo boom. The towed canoe foundered, and we had to cut it free. The small bowsprit foresail kept the boat pointed into the increasingly monstrous waves as we tore through them with hail and rain and lightning crashing from the skies. We had to bale out the boat constantly for many hours. We could see occasional lights on the shore, and we seemed to be making a heck of a storm driven speed for a change. We avoided the shore because the surf would have been big enough to somersault the boat. One of the Spanish guys aboard suddenly remembered his Catholicism and started praying intensively. I invoked Neptune/Poseidon and my servitors silently but with equal fervor.

As the storm began to abate towards dawn, we spotted the lighthouse of Cochin harbor and limped in under sail and oar past the mighty international freight carriers. We were absolutely filthy, and the boat was a bit of a wreck, so we berthed up a small side creek. The next day a small motor launch came aside bearing two locals dressed in immaculate gold buttoned blazers and nautical looking hats. They said they were from the Cochin Yachting Squadron and were looking for a British yacht that they had heard had just come in. I said, 'Yes that's probably us.'

They looked aghast and made their excuses and left. Plainly we were not going to be offered pink gins on the lawns of the yachting club. We swapped the remains of the boat for a few crates of beer and after a few days rest we dispersed on our travels. We took trains to Mangalore and Calcutta and then planes to Thailand and Australia.

177) Are there any results based magic traditions in India that you came in contact with?

Hinduism seems a very broad church with a colorful and extensive pantheon of deities and for some a top god as well, but many consider the top god very remote from this world. It also has many schools of philosophy, yoga, and meditation, and several traditions of holy men. The holy men come in all varieties from those living in caves and begging from villages to those with luxury ashrams and thousands of followers. There seemed a general expectation that any form of religious activity should have a direct benefit of some kind on the material plane and/or result in a materially better reincarnation next time. Thus, the whole system can look like a vast results magic system to the cynical eye, and to the cynical eye so many of the gurus and holy men look like scoundrels.

Whilst we were there, Bhagwan Rajneesh had a huge following amongst young westerners. He basically told them they must have as much sex as possible as a sacred duty and his grateful followers gave him a fleet of gold limousines and a serious messiah complex. One of his Sannyasins joined us for a while on the boat venture and I asked him to tell me what he had learned from his master. He had no answer.

178) What role do you think colonialism played in the evolution of the western esoteric tradition?

Invasions and empires have always helped to spread ideas. The Romans spread Judaism by a forced diaspora and that spread Kabala, they also forced the Egyptian priesthood out of power and into working as itinerant mages. When the Turks conquered Byzantium a whole treasure trove of Classical Greek knowledge was driven westward. Napoleon's invasion of Egypt stimulated a Europe wide fascination with Egyptology. In the nineteenth century, British imperialism paved the way for the assimilation of esoteric ideas borrowed from India and China. Colonialists have usually paid more attention to the esoterics of literary cultures and have borrowed less extensively from the indigenous peoples of the Americas, the Antipodes, the Arctic regions, and sub-Saharan Africa. The Islamic world used to entertain a wealth of esoteric ideas and it helped to preserve them during the European dark ages, but it eventually turned against the 'Greek' style of metaphysical and philosophical enquiry when its territorial expansions ceased.

Basically, ideas about spiritual/occult masters and esoteric bodily systems and their associated practices seem to be colonial era imports into western esotericism which shows little previous evidence of such things.

179) Did esoteric bodily systems and their associated practices bring any new possibilities to the tradition?

The human body has always seemed something of an embarrassment in Christianity, a mere physical vehicle whose base needs and appetites can only impede the immortal soul. Serious christians often practiced various mortifications of the flesh for spiritual reasons.

In some schools of oriental thought, the physical, mental, and spiritual are seen as an holistic unity.

We now recognize the mental benefits of physical fitness and we make little differentiation between the mental and the spiritual.

Thus, western thought and esoteric ideas have moved towards the more holistic view of some oriental systems in this respect.

180) I'm assuming the spiritual/occult master bit is more regressive and harmful than anything else?

Indeed. The idea has little place in western thought where we have come to respect knowledge, ability, and achievement more than mere position.

Spiritual/occult mastership depends almost entirely on position, rather like hereditary monarchy.

The low budget Sadhu positions himself as a holy man by making himself look sufficiently weird to resemble one. The successful Gurus use a great deal of stage management to position themselves as masters to whom others can only relate as supplicants. Few if any of them have anything to offer except their charisma and the supposed 'spiritual privilege' of revering them. Few of them have any unusual knowledge or abilities or achievements to their names.

Educated Asians often take a dim view of holy men today, in this respect western thought has moved eastwards.

Many still worship at the altar of celebrity. Human suggestibility partly results from the psychological payoffs we get from having heroes, exemplars, and role models. Spiritual masters exploit this by promoting themselves as celebrities with unspecifiable talents.

181) What did you learn in creating the hypersphere model that you're applying to creating a quantum model?

Many years ago, I started looking hard at explaining magic in terms of quantum physics. I didn't get very far with it and became diverted into cosmology and the structure of space and time. The conventional theory looked wrong and inadequate. After many years of struggle, I recently completed an alternative and simpler model that fits all the known data just as well. This new model also has a more intuitive feel to it and more satisfying metaphysical implications.

Over the next years I intend to revisit the quantum domain and see if the insights gained about the structure of space and time can elucidate it. Presently the standard quantum model has become a baroque mess, an ad hoc jumble of recipes lacking mathematical or conceptual rigor.

I presently have a mixed heap of possible alternative ideas that do not yet fit together, perhaps something will come out of some combination of them, they include:

A spacetime of three spatial and three temporal dimensions, and the whole metric curved by gravity into a spatial-temporal 3-3 Sphere.

Spins of various of the dimensions of spacetime itself as the basis of all fields, forces, waves, and particles.

Quanta as having some of the properties of hyperspheres. Mach's Principle, as above so below, the quanta and the universe somehow mutually define each other's properties.

Matter 'particles' and 'anti-particles' as the points where waves which travel right round the universe interfere, a sort of hyperspherical extension to Milo Wolff's wave structure of matter.

Energy 'particles' as 'particle-antiparticle' exchanges with the antiparticle component traveling backwards in time, as in the Transactional Interpretation of quantum physics.

The appearance of entanglement and superposition arising from retro-causation at lightspeed.

A relationship between neutrinos, photons, and gravitons.

The hypothesis that something out there might know more than we do, and that it might respond to the name of Azathoth.

182) So, you're trying to get in touch with Azathoth then, or in talks with it? Did it help you on your cosmological model?

As a Chaoist I have the liberty to entertain mutually contradictory hypotheses. One hypothesis says that Azathoth represents something in my subconscious that can better understand quanta than my conscious mind can. Another hypothesis says that the universe must contain extra-terrestrial intelligences that know more about quanta than humans currently do, and that information can pass across space somehow.

I decided to follow H. P. Lovecraft's lead on this and imagine the possibility of receiving information from what he called the Elder God Azathoth that presides over the Nuclear Chaos.

Of course, I don't expect it to speak English, but I hope it might think in geometry and math of some sort.

Either way, the practical application of each hypothesis remains the same. I fill my conscious mind and memory with as many conventional, speculative, and unconventional ideas as possible, paying particular attention to anomalies, and then allow my Subconscious and/or the Elder Gods to prompt new imaginings which I afterwards subject to rigorous analysis.

Some people would argue that neither the Subconscious nor The Elder Gods really exist.

I remain content with the magical idea that imaginary phenomena can have very real effects.

Magic is what you use when you have exhausted the possibilities of common sense.

It seems ideally suited to attempting the impossible.

For the cosmological work I conceptualized the source of inspiration as Yog-Sothoth, another of Lovecraft's Elder Gods. The resulting Hypersphere Cosmology appears to have done what cosmologists consider to be impossible, making a non-expanding model of the universe that resists falsification.

183) Do sacrilege, heresy, iconoclasm, and anathemas have a basis in older traditions?

The Zen Koan that runs "If you meet the Buddha on the road, Kill him", springs to mind. Taoism also values the contemplation of opposites to socially acceptable ideas. Some sects of Hindu sadhus and holy men deliberately break as many social taboos as possible as a matter of mystical policy.

The Gnostic sects of the first few centuries AD believed in a hierarchy of Archons stretching between the earthly and divine realms. Some Aesthetic Gnostic sects attempted to avoid and transcend the specific 'vices' associated with each of the Archons (which corresponded to some of the Olympian gods and goddesses), other Sybaritic Gnostic sects sought to fully indulge such vices and to progress upward through them.

Christendom developed an underground cult of demonology and demonolatry, if not outright satanism, in the Grimoire tradition.

People only invest serious emotional commitment to highly irrational beliefs that come shadowed with great potential for doubt. Virgin birth, resurrection of the dead, life after death, a god that loves everyone personally despite dealing many of them a lot of horrible life experiences, all these contra-intuitive ideas demand considerable investments of faith and willful stupidity to maintain.

Breaking a belief liberates free belief that we can invest elsewhere, perhaps for a better return. This underlies the contrarian practices and contemplations of some eastern schools of enlightenment. It has also inspired many magical traditions in terms of their beliefs and practices.

Magic has always had a rather difficult relationship with religion. Followers of religions have always expected their religions to create some 'magical' effects in terms of miracles, blessings, and benefits.

Magical practitioners have often borrowed religious theories and language to explain their activities. Religious authorities have invariably opposed any form of magic not sanctioned by themselves. This probably encouraged magicians to find inspiration in inverting or contradicting religious ideas.

The revival of Witchcraft in the 1960s and 70s was basically a re-invention rather than a revival and it strongly identified itself by its aggressive rejection of Christianity and all its ideals, however it has since become considerably tamer as it became more and more commercialized.

Chaos Magic partly defined itself by its opposition to Crowleyan Thelema which had defined itself pretty much as the opposite of all three Abrahamic monotheisms. This didn't make Chaoists into monotheists incidentally; Pagan-Atheist-Parapsychologists might perhaps better describe us, or at least some of us some of the time.

Interview 6

184) When did the idea of the magical diary enter the western tradition and where did it come from?

I suspect the tradition comes mainly from Crowley. He kept voluminous diaries, with, I suspect, an eye to publication and posterity. His diaries consist of a mix of yoga style exercises and magical workings, mainly Invocations.

185) When did you start yours?

I have adopted the practice from time to time, but I have not found much value in recording routine exercises performed, rather I have tended to just record magical workings that went well, for future use or for use by others. In many ways my books constitute my magical diaries.

186) Did it take a while for you to develop a strict habit in that area?

Crowley lived the leisured life of a wealthy gentleman, indulging the habit of extensive diarism and meditation would have proved easy for him. If you try it in the context of a busy working lifestyle it can become an ordeal. I have done it formally for a few months at a time but in the end, I realized the diary ordeal had become more of an issue than the practice.

187) Do you still use that tool?

No. However I always record the fruits of my magical workings and my imagination.

188) Do you read back on the old stuff?

No. When I reviewed the routine exercises material it seemed no more interesting than a record of miles clocked up on a static exercise bike. I discarded such records, but I kept the material on my successful Invocations and Rituals until I had written them up formally for their future use by myself or others.

189) What's the biggest insight you've gained from that habit?

I have learned to value Imagination over Will. You can only muster what seems like 'willpower' by using or engaging the Imagination to persuade yourself of the value of doing something or refraining from doing something.

I always remember what my imagination creates, and I usually translate that into a piece of writing or mathematics, or a magical artifact, or a sculpture. My writing, artifacts and sculptures constitute my real magical diaries as I eventually realized.

The long hours spent in the Gnostic Pentagram Ritual, Pranayama, and the Death Posture did seem to make my imagination work better afterwards precisely because they suppressed it during the practice.

Visualization practice not only strengthens the imagination, but it also teaches control of attention in a very direct way. Most people have rather poor control over their attention, they either cannot maintain it for long or they cannot switch it away from dysfunctional activities.

I prefer to set time aside on alternate days for practice, usually a mix of movements and visualizations. I used to do this before sleep but since retiring from commerce I have tended to practice in the mornings, from 10 am onwards. As an Owl rather than a Lark I tend to sleep late and long. Nothing good usually comes of anything done before 10am in the morning in my experience.

190) What about dream diaries? I assume they also come from Crowley?

I cannot recall Crowley making much of his dreams. I suspect the idea comes from early psychoanalysis, particularly Jung.

191) Do you keep one of those?

I have gone through periods of keeping a dream diary. It often becomes a rather maddening enterprise because part of the function of dreams seems to be 'exorcism', we often mash up the trivial memories of the day in order to defuse and forget them. I often found myself going over them in the morning and thinking, I know why I dreamed that, now I need to forget it all over again.

192) Dream work seems like it takes a lot of consistent practice, how long did it take you to "master" dream work and what does that look like?

I started trying to find my hands in dream and became involved in strange concentric dreams where I started dreaming about dreaming that I had found my hands, and even dreaming that I had filled in a dream diary when I had not.

193) Do you just live a whole other conscious life in your dreams at this point?

In general, the more conscious I become in dreams the more likely I become to awake. As soon as I try to shape the dream, I wake up. However, I have woken from several 'telepathic' dreams in which specific real people said something of importance to me which I later confirmed by telephone.

I have increasingly found the 'Liminal State' between waking and sleeping my most effective 'Alternative State' for creativity, imagination, and inspiration. I make a point of keeping a notepad by the bed so that ideas do not get lost when I fall asleep or fall back to sleep. I have even cracked challenging bits of algebra that I could not do whilst fully awake. I try to enter sleep and to exit it as slowly as possible in order to maximize time spent in the liminal state.

For me, the main value in dreamwork seems to have been the better access it has given me to the liminal state.

194) What sort of sculpture do you do?

I mainly do mythological and fantasy sculptures and I also make my own magical instruments. I generally prefer materials that can be worked by both addition and subtraction (as this allows for changes of mind) so I rarely try to carve wood or stone except in the simplest ways, as you cannot put material back on if you carve off too much. As part of my Druidry studies I have done a group of local Celtic deities by applying two-part resin putty over steel wire and then working it with tools after hardening before applying verdigris with acrylics to produce an aged bronze look. I have done a group of Elder Gods in oven hardening resin putty to show my graphic artist what I wanted for the Epoch (I never got the hang of two-dimensional drawing). For the local community gardens, I made a set of large wall plaques in deep relief showing Sun, Moon, a local Goddess, and Green Man faces and a vertical sundial with a metal Saturn figure on it, these I did using experimental cement and compost mixtures. I also did three life-size sculptures of classical deities using concrete plastered over shop dummies. I have also dabbled in casting small figurines in solid tin which you can melt on a gas stove. As you can guess, I basically like to dabble with unusual materials to make unusual unique objects, primarily because I wanted such objects to exist for me. Perhaps the most unusual consists of a two-foot-high wooden carving of the goddess Apophenia mounted as a chaotic torsion pendulum with a contra-rotating malachite globe set in a four-foot-high wooden dowel tetrahedron above a set of divination discs. My home lies littered with unique self-made artifacts which mean more to me than anything I have merely bought. I rather enjoy making fantasy starships out of found materials because there are no rules about what they could or should look like except that they should each have a distinctive aesthetic and stimulate the desire to give them an imaginative backstory.

195) Which of these mental states does sculpting most resemble?

I go through a whole gamut of states and emotions when making artifacts. Anger and frustration when things go wrong, anger often seems to promote a creative breakthrough - I do not reject it, elation when things go well, calm satisfaction during the fine finishing of a piece that went well. Every piece starts as an experiment, I rarely have an exact vision of the final form when I begin, so the result always comes as something of a surprise.

196) How important is mastery of no mind to results magic?

'If You Can Stop Thinking, You Can Do Anything.'

I reckon that remains the most powerful and dangerous phrase ever uttered.

Doubt, fear, hesitation, and procrastination hold people back from doing so many things (often for good reason, it keeps them alive), but if you can let go of it you can at least attempt anything, even seemingly impossible things.

The trick of course lies in deciding what you want to do beforehand and then going for it unreservedly without further conscious deliberation.

One of the main virtues of magical rituals for example, lies in occupying the operator in a performance to make something happen that does not leave space for further thought, distraction, and doubt to arise once it commences.

Even in ordinary life the not-thinking principle proves invaluable. Many people fear public speaking because they think they will freeze up trying to think about what they are speaking about whilst speaking, instead of simply speaking. The trick here lies in knowing your material beforehand and not caring how you express it. Take no more than five topic keywords in on a prompt card for even the longest of speeches.

Many people shy away from writing at length and start overthinking the first sentence. When people ask me how to become a writer I say, do not try, if you can speak you can write, take it from there, your own not-thinking mind will come up with plenty of material if you let it.

197) Do you feel you've mastered that?

I doubt that anyone ever fully masters it in the sense that they always act spontaneously, unreservedly and without due consideration. Such people would appear crazy and a danger to themselves. However, I have developed some skill in turning it on and off.

198) If so, how long did that take?

It seemed to gradually grow on me, most of my skills seemed to improve as a side effect of practicing magic, but I have heard the same said of many mind-body activities, particularly martial arts, perhaps all these impart the ability to act confidently without too much conscious deliberation.

199) How long did it take you to get proficient at visualization?

I think everyone can visualize things momentarily, but it took me a while to create stable and persistent images.

200) Any tips on getting good at that?

Visualization can take many forms, visualizing things in the mind's eye - for example imagining a triangle hanging in mid-air and maybe rotating it, or closing your eyes and 'seeing' a triangle as bars of light, or perhaps projecting an hallucination of a triangle on to the environment with your eyes open. It does not matter what form your visualizations take so long as you can make them persist in the desired form. Pentagrams prove slightly more challenging. Rotating imaginary dice and keeping all the numbered faces in the right places can provide the next challenge. Such exercises strengthen the ability to visualize useful things like sigils or spells, or 'spirits' or desirable outcomes in physical reality.

201) So, the Irish president has a role very similar to the constitutional monarch, do you think it's just as effective? There is nothing missing from the equation because he isn't "royal"?

Power always relies on a mixture of a monopoly of force and a fair bit of trickery. Leaders must use plenty of stage management as well as threat of force or force itself. The Irish Presidency and the British Monarchy depend entirely on the arrangement that both they and the populace will pretend that they are in power and have ultimate veto over the elected government. Plus, both nations have independent judiciaries and in Britain at least we only allow posh people to rise to senior military rank. All these checks and balances have evolved pragmatically to prevent revolutions, military coups, and elective dictatorships. So far so good. The ancient Greeks concluded that the ideal form of government would consist of a mixture of monarchy, aristocracy, and democracy, although they rarely achieved it.

202) What about the royal family's wealth? Is it large enough to impact their influence or are they basically just some rich people with a government role?

We still consider that the Monarchy should maintain a lot of bling to stage manage their assigned position of pretend power. However, some people question the amount. The greatest threat to the British Monarchy seems to lie in the all too human behavior of the minor royals, of which we have too many. We pay them to pretend to be better than us.

203) Let's say you were to define anarchism more broadly as an absence of coercion and class division?

People are not born with equal abilities and they will sort themselves into classes and hierarchies of some sort or another unless you apply some sort of coercion. I doubt that you can have any sort of society without at least some coercive facilities. At least we have realized that the instrument of money usually coerces better than clubs and spears. Society depends on agreements about monopoly of force and law, and nowadays also on agreements and shared beliefs about the value of currencies. Historically, religions have contributed a lot to the trickery of power. Secular and religious power have usually worked together to coerce the desired behavior from the populace. All theology, all the doctrines of sin and virtue, heaven and hell, all ultimately come down to politics – how people should behave towards each other and the demands of 'the state'.

As a species we do not live as solitary creatures looking out only for ourselves and meeting only to mate. We live as social animals that can kill each other ridiculously easily in disputes over mates, resources, and lebensraum. Cooperation comes at the price of collective agreements and collusions about coercion, divisions of labor, and hierarchies. We shall forever argue about the details of how to do this, but I do not think we can do without such things in principle.

The only thing worse than democracy is everything else we ever tried – Winston Churchill.

However, I do wonder if Chaocracy might work even better. Keep the civil service in place but select a parliament randomly by lot every few years. Pay them handsomely to turn up, debate, vote, and govern, but have some mechanism to punish them ferociously if they try to line their own pockets further. This would at least relieve us of the burden of reluctantly voting for the least worse seeming and the least seemingly dishonest career politician at each election.

204) Here there was some buzz for a while around something they called star voting but basically you get to vote for your top 3 or top 5 choices instead of just one candidate or whatever. Sound interesting at all to you?

When the conservatives in the UK had to form an alliance with the minor liberal democrat party, the liberal's price was a referendum on alternative votes. The country rejected this largely on the grounds that it would probably have given the small liberal party a disproportionately large number of seats. People in the UK tend to vote negatively, party membership remains extremely low, people vote to keep out the party they dislike most.

Some countries operate a None of the Above system. If that gets the most votes, the election is repeated with none of the previous candidates eligible to stand again.

Part of the momentum for Brexit was maintained by the system of proportional representation used for the sham European parliament. Many constituencies sent a UK Independence Party candidate as all constituencies sent the top two or three vote winners. The UKIP representatives in the sham euro-parliament spent all their time there making a splendidly loud nuisance of themselves.

205) Was there anything about the EU that you were initially hopeful about?

In 1975 the UK conducted a referendum on continued membership of the EEC or 'Common Market'. I was only 22 at the time, barely interested in politics, and had not registered to vote anyway. I had just left a London college where the student union politicos sent a new final demand to the United Nations almost every week. I found them very irritating.

The far right and the far left formed a strange alliance to oppose continued EEC membership. The far right said it would lead to an erosion of national sovereignty, the far left said it would lead to domination by big business. They lost, but they were both eventually proved correct.

I did not take much interest in politics until it began to impact my small business more than a decade later as I tried to grow it into a medium sized business. The EEC had by then mutated into the EU and developed a federalist political agenda that went far beyond its original brief of economic cooperation.

The EU basically outsourced its regulatory culture to big business and big business jumped at the opportunity to regulate small business to prevent it competing with big business.

The more I read and heard about the EU the less I liked it. The UK had been seduced into giving away a lot of control over its own affairs in return for easier terms of trade. The control of British affairs seemed headed towards control by a largely unelected Synarchy of eurocrats and its vast bureaucracy.

I joined the United Kingdom Independence Party and became a local treasurer and activist, it seemed a rather hopeless quixotic gesture at the time, but in the end, we got Brexit.

If you have reasonable opinions, I cannot recommend joining any political party in the UK. Most party members and activists have opinions far more extreme than their party publicly advertises, and many of them have strange ulterior motives.

The fight for Brexit got ugly and dirty at times, the country became very polarized over the issue, and the result was uncomfortably close and a shock to pundits and pollsters. However, it now seems highly unlikely that the UK will ever seek to rejoin the EU in its current corrupt, inefficient, and undemocratic form.

206) Could Chaocracy work for a town or city? Like as a proof-of-concept experiment?

The ancient Greeks experimented with Sortition or Stochocracy, the selection of people for positions of power and influence by lottery. Britain uses it to select juries in courts, sometimes it is used to fill advisory committees. It certainly has the advantage of ensuring a representative selection of both sexes and all classes, plus it reduces factionalism and the role of organized political parties. Of course, we would end up with five percent of idiots in government, but that would probably represent a reduction in what we usually get.

207) In the news here, Brexit is painted as being part of the rise of nationalism in "the developed world", Donald Trump, racism, etc. Do you think that played a role in the political support it got?

Globalization has had many negative effects upon the less wealthy in the developed world. Immigration depresses wages and inflates property prices; cheap imports have eroded skilled manufacturing employment opportunities. The offshoring of capital has starved industry of investment. A great many people in the working and lower middle classes have seen their prospects diminished by Globalization and they voted for Brexit and Trump in consequence.

The British seem more culturalist than racist. We accept those who accept our culture. Many in the second generation of West Indians, born in the UK, often tended to reject British culture and this led to a lot of trouble, but the third generation seems to have settled in better. The Hindu Indians from Uganda settled in and quickly opened their own businesses in this nation of shopkeepers. However, the Moslems have, in some cases, maintained Islamic attitudes that remain diametrically opposed to British values, and the resulting problems may take another generation or two to go away.

We have one of the most ridiculously overcrowded nations on earth and it has begun to dawn on many of us that any further growth of population will have a negative rather than a positive effect on quality of life here. It may seem good for business in the short term, but we are running out of countryside and farmland to concrete over, and the environment continually deteriorates. Regaining control over immigration motivated many to vote for Brexit.

208) Do you think there is a rise in nationalist attitude in the UK?

Yes, I think closer contact with Europe has served to illustrate the big differences between us and them. We do not want a French style justice system, or a German style system of social control, or southern European levels of corruption, or Belgian levels of bureaucratic screw up.

Nationalism has really taken off in Scotland where a substantial proportion of the electorate seems to want to separate itself from England, and in Wales the nationalist party has made considerable inroads.

Patriotism went out of the window after WW2 in Britain. We won an almost pyrrhic victory that cost this country its empire and bankrupted it. In dismal postwar Britain, patriotism became deeply unfashionable, and contemptible to the political liberal left, it became associated only with the far-right fringe. Britain has a continuing identity crisis; we cannot decide whether to feel proud or ashamed of our history.

Part of the motivation for throwing in our lot with Europe lay in the feeling that Britain was finished as a world power and a great independent nation. Perhaps that feeling will change now that we have regained independence.

209) It seems like being proud or ashamed of one's national history has played a large part in the political division in both our countries recently, do you see a best possible or most likely outcome to the problem of national disunity?

"Who controls the past controls the future: who controls the present controls the past." George Orwell - Nineteen Eighty-Four.

We have always weaponised history and fought over how to present our history to ourselves. Even Shakespeare's historical plays contain a lot of pro-Tudor propaganda. The ancient Egyptian inscriptions recording a glorious victory at the battle of Kadesh quite likely conceal a defeat.

To me it seems important that we do not censor or destroy historical records and interpretations that have become unfashionable or embarrassing.

It seems that you can only find opinion free objective history about archaeological events, and even then, the authors often had some axe to grind, to which later historians have often added their own 'interpretations' based on their own political feelings.

210) Do you think the solution to our environmental problems is more scientific or more political?

'Sustainable Growth' looks more and more like an oxymoronic term. Yet almost all politicians still advocate perpetual economic growth and very few advocate population limitations.

There seems an underlying hope that 'Demographic Transitions' will occur everywhere when everyone considers that they have enough wealth and security to refrain from producing offspring at more than the replacement rate. Yet this planet does not seem to have sufficient resources to permit this even for the current human population at current western levels of consumption.

Politicians will only start advocating less when it becomes painfully apparent to their populaces that more means less in the short term; when the environment starts kicking us much harder than we are kicking it. I suspect things will have to get a lot worse before that begins to happen.

The technical and scientific fixes we currently consider may buy us some time. Genetic engineering of food crops could reduce some of the demands we place upon the land. Renewable energy sources could reduce carbon emissions. Ingenuity could find substitutes for increasingly scarce resources and minerals.

At present, schemes for mass migration to other planets, liberating unlimited power from hydrogen fusion, and geoengineering the climate non-catastrophically all seem equally unlikely.

I feel that I have lived at an extremely fortunate time in fortunate places, perhaps at the zenith of a civilization that will not prove sustainable.

The future looks set to become a close-run thing.

211) When you say the first real 'occult revival' began in the first and second centuries, do you mean in the sense that it was referring back to prehistoric shamanism?

Magical ideas and magical thinking always seem present to some degree in all human cultures. A 'magical revival' occurs when this form of thinking comes to the fore with new ideas and has a more than usual influence on a culture. Such revivals usually accompany eras of social upheaval and change. Historians and occult theorists may well argue endlessly about the significance of various resurgences of magical ideas but three 'occult revivals' seem to have had a particularly seminal impact on western magical thinking.

In the first and second centuries AD the Roman-Hellenic world had incorporated a huge variety of peoples and ideas into its civilization. The classical Greco-Roman Olympian Paganism attempted to absorb Greek Philosophy, Judaic Monotheism, and Egyptian religious and magical ideas, to mention just the more well-known influences. In this it failed. Instead, new cults and sects and ideas proliferated, picking, and mixing from the cornucopia of a dynamical spiritual and occult marketplace. Hermeticism, Gnosticism, and Kabala all date from this period as does the Christianity which eventually supplanted the Olympian religion.

Some might argue that significant occult revivals occurred during the Renaissance and in response to the Reformation, but these seem largely a revival of the same ideas that arose in the first revival and had gone underground under the influence of Christianity.

The second major western revival began in the late Nineteenth century when oriental ideas, mainly from India and China, were brought in through colonialism and empire and mixed with the old western cannon and a touch of Enlightenment thought and Industrial Revolution science. Thus, we ended up with Theosophy, Spiritualism, Eliphas Levi, and the Golden Dawn.

A third significant western revival began in the later years of the Twentieth century. From the late sixties onwards, massive social changes occurred. Many people in western societies became wealthier and freer and better educated and many of the old certainties went out of the window to be replaced by new uncertainties and opportunities. Imported oriental and ethnographical ideas from all over became available to the masses rather than to just scholars and elites.

A cacophony of cults, sects, and pseudo-ancient traditions sprung up, but the third revival also led to the growth of a Meta-Perspective that owes a lot to science.

This Meta-Perspective finds perhaps it first explicit expression in the works of Schopenhauer 1778-1860. He had widespread interests in esoteric, parapsychological, occult, and magical phenomena and in religions, but he did not accept any of the traditional spiritual ideas and beliefs used to explain them. Rather he considered that all such phenomena arose through the belief, will, and imagination of the participants in magic and religion.

Today, in the aftermath of the third magical revival, a high proportion of western practitioners and 'believers' tend to opine that they invest belief and effort in their magic or religion because it works for them (perhaps by some strange scientific mechanism), rather than because it's absolutely true for some 'higher spiritual' reason. Having either killed all the gods or recognized them as parts of our own psychology, we can all become wizards and sorcerers.

Whether or not the Animism and Shamanism of our hunter-gatherer ancestors constituted a zeroth magical revival remains a matter of debate. Arguably most of the main themes of magic have their roots in Animistic and Shamanic practices.

212) Since the Knights of Chaos came out of the online college, I assume you were coordinating your work online?

Yes, for a time the KoC conducted a series of successful campaigns with a substantial number of participants worldwide working towards mutually acceptable objectives.

213) Did that teach you anything new about group magic?

After a while we ran out of mutually agreeable objectives. Any intervention in world affairs will have its upsides and its potential downsides. Having picked most of the low hanging fruit with mainly obvious upside consequences, we could achieve no further consensus, particularly on matters of other people's religions and politics and economic growth in general.

214) Did it hinder or help anything for the group in general?

The internet seems no place for the sort of extended philosophical and ethical debates that might have led to the forging of fresh consensus, for that you need face to face meetings.

215) Are there any magical entities that have been helping you for a while?

In the first half of my career, I took considerable inspiration from Baphomet as portrayed by Eliphas Levi and I invoked it many times. It embodied my interest in my sexuality, my desire to act dynamically and energetically in many fields, and a certain daemonic antinomianism.

After a while I graduated to Ouranos, a god form of pure magic that for me completed the seven traditional planetary god forms into an Ogdoad, and to which I assigned the 'imaginary' color Octarine. This for me went hand in hand with a period of extensive magical research.

Later I progressed to invocations of Apophenia, an entirely new goddess with a partly Ouranian nature but more generally specialized in finding the hidden connections which may or may not exist between ideas and between phenomena. This for me goes hand in hand with a broadening of my research into cosmology and quanta in search of new metaphysics.

Apophenia has led me towards a neo-Lovecraftian conception of the Elder Gods as possible extra-terrestrial sources of scientific and metaphysical knowledge. In particular, Yog-Sothoth for knowledge of time and space, and Azathoth for knowledge of mass and energy.

Of course, I have had other dalliances along the way, Jupiter for business success, Eris as a female version of Baphomet, and Vulcan for my works of artifice.

216) Has working with these sorts of beings broadened your view or definition of consciousness?

I regard consciousness as an activity that the brain performs, and that activity includes creating the illusion of a 'unitary self'. Despite that illusion, our brains can execute an enormous variety of performances, we have worlds within us.

217) What do you mean by god-forms?

For me, calling forth a god-form means reprogramming the brain and putting a mythological label on the reprogrammed area to enhance future access.

218) Do they start to take on more of their own character or more autonomy with time?

Yes indeed, and to avoid obsession a magician should have several god-forms available and preferably an entire pantheon for a full life.

219) Do you think acquainting yourself with entities that may or may not exist outside yourself, gives you an advantage in recognizing entities or intelligences that do exist outside of you? Like I've often wonder if an alien dropped down in front of me, if my brain would even be able to make sense of, or register what was happening and if my brain did register it, if I would believe my eyes or just loose it.

I have seen extraordinarily little evidence for 'gaseous vertebrates' or discarnate spirits. Sapience always appears in embodied form. However, some form of psychism or telepathy may perhaps enhance whatever you invoke.

Anything clever enough to get itself here from another star system would probably have the ability to appear as anything it wished, or to remain completely invisible to us.

220) How long have languages for purely magical reasons been a thing?

The Norse Runes seem to have evolved as an imitation of Latin script that users could carve or chisel onto wood or metal or stone, hence the lack of curves in it. Initially the Norse peoples seem to have used Runes sparingly as magical ideograms. Thus, a weapon for example might have runes for the gods Thor and Tiwaz and maybe others to represent the owner on it. Later it became used for whole words, and then phrases and discursive writing.

221) Are they fairly common in magical traditions?

All hieroglyphs, letters, words, and symbols represent thoughts, and they have the astonishing property that they can preserve thoughts and communicate thoughts to others across time and space. The ancient Egyptians considered that Thoth, the god of magic himself, had invented writing.

The purely magical use of language and writing depends on using it in obtuse symbolic form so that it evokes subconscious rather than conscious thought. Thus, magicians often prefer to express spells and incantations in scripts and words that make no immediate sense to their conscious minds, but which activate their subconscious minds because on some level the mind remembers the meaning transcribed into the weird scripts and words.

222) You've just created the one?

Initially I started transcribing incantations and invocations into Enochian, a strange 'language' developed by Dee and Kelly in the 16th Century. It has a fine resonant sound to it but only a rather limited lexicon. It seems to have no historical precedent, Dee and Kelly appear to have conjured it up through psychic and imaginative means that appear to have involved the use of tables and squares of letters and symbols.

After a while I realized that modern magicians needed something more flexible that allowed for the creation of a magical version of any word or concept. At the time I was doing extensive research invocations of Ouranos, and the Ouranian-Barbaric language came out of this. To obtain the Ouranian-Barbaric words an operator performs an invocation to Ouranos whilst whirling on the spot Dervish style and focusing on a palm sized Ouranos pentacle held in the left hand. After a while, a scribe calls out a word for transmogrification and the operator slams the pentacle down several times as inspired onto a board of squares each bearing one of the consonants which the scribe duly records. Afterwards participants meditate upon a suitable vowel pointing. XQL for example became Xiqual, pronounced Ziqual. One scribe and several operators working in relays can produce quite a few words in a session.

223) What makes for a most effective magical language?

Bronowski's anthropological observation that such utterances should have a high level of weirdness holds true. They should not stimulate ordinary conscious discursive thought. I can remember two longish incantations, one in Enochian and one in Ouranian-Barbaric, subconsciously I know what they mean, I wrote them, but I would have to look up the English from which I transcribed them.

224) Or are there different qualities that work best for certain applications?

Ouranian-Barbaric seems to work well for most purposes, you can always adjust the tone and emotion with which you incant it to suit the ambience of the work.

225) Are any or most of them complete enough that one could carry on a conversation or would that be counterproductive to their working on an unconscious level?

If you became fluent enough in a magical language to conduct a conversation it would cease to work well as a magical language. The O-B lexicon consists mainly of declamatory rather than conversational words and concepts.

Interview 7

226) What are your thoughts on universal basic income?

I remain opposed to anything that might ensure the survival or reproduction of anyone who contributes nothing to the human enterprise. Work or die.

Work here can mean doing almost anything that at least some others appreciate enough to cooperate in feeding you.

People have no intrinsic value to me, for me it's what they do that counts. Paying people to do nothing sounds like an exceptionally bad idea.

227) How about nuclear power?

I doubt that Fusion will ever prove worthwhile, we have tried for nearly eighty years to make it work with little success. Fusion power is always just twenty years away. A cubic meter of the sun's core is thought to produce only about one watt. That may sound surprising, but it has burned very slowly for billions of years and it contains an almost unimaginably large number of cubic meters. We would need to speed up the hydrogen to helium reaction by a factor of a millionfold and somehow contain the reaction and somehow extract heat energy from it.

The math seems against any of the schemes currently under investigation. I think we need more fundamental research before throwing any more money and effort at any of them.

We can tame Fission but at a terrible cost because we can never properly decommission the reactors or render the wastes harmless. The true long-term cost of power from Fission seems incalculably high.

We now understand that if we do not leave the rest of the fossil fuels underground and unexploited then the climate will change catastrophically.

If humanity could act with collective good sense, it would immediately go for all possible renewable forms of power; solar, wind, hydroelectric, wave, tidal, geothermal, etc., and force itself to adjust to considerably lower living standards.

Humanity has become fat and bloated on fossil carbon that has fueled an unsustainable growth in population and consumption, and which now threatens to render the planet uninhabitable to us and many other species.

228) Can virtual reality get to a sufficiently immersive point as to be ritually useful?

'Astral Magic' as I have defined it depends on creating an artificial environment by visualization, so with closed eyes, laying or sitting down, the magician imagines a whole ritual set up complete with all desired instruments and equipment and imagines going through a ritual procedure in fine detail. Some magicians choose imaginary forest glades or underground vaults, some even use spacecraft. Such imaginary virtual realities take a considerable time to create and stabilize in the imagination, but such a facility seems far more useful than anything I have heard about goggles and headsets and movement sensitive gloves.

229) Sorry, this is a vague one, but I'm interested in everything you know about ritual dance, its history and rhythm. The history of rulers trying to control or outlaw dances fascinates me, particularly here in America.

Dances tends to become outlawed either because they offend sexual morality or because they represent cultural or religious practices that someone in power want to suppress.

The historical list of groups accused of 'Dancing with the Devil' seems endless. Some modern occult groups embrace this principle and dance shamelessly with the Horned God or Baphomet or some similar symbol of rebellious abandon. In general, I prefer to dance naked.

230) What sort of ritual dancing do you have experience in and or prefer?

I prefer to use dance for self-expression and gnosis, rather than for communication or entertainment or social purposes. The Mass of Chaos Ritual which I devised and frequently participate in calls for a wild circle dance, but participants do this without reference to others, apart from avoiding collisions. Dervish style whirling has often proved a useful technique for enchantment and divination, particularly as you can hold a sigil or spell in the palm of your left hand as a focus for balance and concentration whilst spinning widdershins.

231) Is faith a useful concept in any context?

It always seems a better idea to acknowledge that you have faith in something rather than that something 'Is True'. Faith acts as a powerful but double-edged sword, it can carry its bearer past doubt and difficulty to succeed against the odds but it all to easily leads to disaster if it passes too far beyond sense and reason.

That abominable little thug Napoleon Bonaparte said something to the effect that, "in war, moral power is to the physical as three parts out of four.", although at another point he apparently conceded that 'God is on the side with the most artillery'. Both he and Hitler won cheap and quick early victories, against larger forces, using armies that adopted faith in their ideas with great enthusiasm. Yet their own faith in themselves and their follower's faith in them, led to a bad end for all concerned.

Faith in life after death seems one of the most widespread and peculiar of all human beliefs. Historically it has served three purposes, it has supplied a social control mechanism via conditional immortality (behave as instructed now, become rewarded or punished later), it has supported social inequality (the meek shall reap their reward in heaven), and it has also proved an invaluable incentive to risk death in war. It would seem an interesting statistical exercise to find out whether belief in an afterlife increases or decreases life expectancy. I suspect that most of the life expectancy increasing innovations have come from people who did not believe in an afterlife.

Personally, I find the most functional form of faith lies in slightly overestimating your own abilities. This stimulates incremental growth and avoids major disasters.

232) Is there such a thing as truth?

Well, it certainly exists as a concept. In practice it tends to mean a description that accords well with observations, or with prejudices.

Religions usually take care to make their 'truths' as unfalsifiable as possible. They endeavor to make claims that nobody can easily disprove. If anyone gets close, they just move the goalposts.

Science, in principle, takes the opposite view. A scientific truth should remain open to falsification. Confirmation merely adds to confidence or faith in the idea but if the idea remains potentially falsifiable then it never constitutes an absolute truth.

I hate it when cosmologists opine that 'We Know the Universe is Expanding and composed of 70% Dark Energy and 25% Dark Matter'. They do not. They merely have observations that they can interpret to mean that. Other interpretations remain possible. New observations may yet overturn the hypothesis. Nevertheless, it seems highly unlikely that any observation will overturn the hypothesis that the Earth consists of an approximately spherical astronomical body.

In the quantum realm which supposedly underlies the whole of reality, many people suspect that the descriptions we have so far developed are little more than mathematical formalisms created to give some predictive power, but which tell us nothing about what 'really' happens at that level of reality. Even Murray Gell-Mann who came up with the Nobel Prize winning quark theory of matter, often wondered whether quarks 'really' exist or whether the quark idea was simply a mathematical trick that could explain some of the observed behavior of quanta.

For me, the best Magical Truths consist of conditional pragmatisms:

Enchanting Long and Divining Short usually gives the most useful results.

Nothing has ultimate truth, anything remains possible, but the results may displease.

Belief usually works best as a tool rather than as an end in itself.

Interroga Omnia – Question all things. (But not all the time)

233) I guess Celtic magic would be the local flavor in your area. Is there anything unique you learned from that tradition?

The iron age peoples of pre-Roman Britain had no writing, so we only have archaeology and what the Roman invaders wrote to give us any clues about Celtic culture. The Romans did not invade Ireland, but they did bring Christianity and literacy later. Much of what Roman scholars wrote of pre-Roman Britain and what Christian monks wrote of pre-Christian Ireland looks like propaganda rather than anthropology. Iron age Celtic cultures seem to have had no interest in Stonehenge and all the megaliths left by the previous late Neolithic cultures.

Thus today, Celtic Esoterics has become a romantic syncretism based on a very thin historical knowledge base and a large admixture of more modern occultism. In other words, modern practitioners basically make it up as they go along using the few bits of surviving myth preserved in Romano-British/Welsh and Irish Christian writings, plus a limited amount of archaeological iconography. I do attend a local Druid Grove that takes a very eclectic and a somewhat a-historical approach because we have so little reliable information about what the ancient Druids did. In effect Druidry and Celtic esoterics have become yet another 'flavor' of modern occultism along with Wicca and Neo-Paganism in general.

234) Do you think there is anything we could do to internet configuration or administration to counteract its negative cultural influences?

I do not interact with social media. I maintain a personal website where I present my ideas and I respond to emails sent to it. This works fine for me. There are no images of me on the internet as I do not wish to be mobbed on the street by those whose metaphysical, cosmological, religious, and political ideas I have contradicted.

I also use the internet for research, and it seems as useful as having an enormous library of reference books on any imaginable subject - but compiled by an indiscriminate librarian.

A lot of people complain about the negative effects of social media, yet they continue to use it, despite in most cases knowing that it is designed to be addictive and that it contains a lot of lies and disinformation and that it harvests personal data for sale. As the saying goes – if you are not paying for it then you are the product.

The internet seems to multiply human intelligence and stupidity in equal measure. Do internet gamblers not realize that they play not against chance but against algorithms designed by professors of psychology and mathematics to relieve them of the maximum possible amount of money?

We have accepted the principle that any potentially dangerous product should carry appropriate safety warnings. Maybe the internet should have to flash up a general-purpose fraud, bullshit, and disinformation warning – may contain nuts, every time anyone switches it on.

235) Let's imagine that virtual reality gets really good, all the senses super engaging etc., etc., which I'm assuming it will one day, what *could* the positive potential be?

It will open up enormous possibilities for the pornography industry. Tele-dildonics could allow people to have 'intimate' physical contact remotely. Computer graphics could even remove the need for human actors completely or allow for the insertion of anyone's image and mannerisms into the entertainment. I guess we could also have bespoke snuff movies where you can murder someone you hate in a suitably gruesome manner.

I doubt that such virtual enactments would increase the probabilities of desired sexual liaisons or hoped for deaths in the real world more effectively than conventional magical procedures.

236) There has recently been a string of strange cattle mutilations a few hours east of me, no tracks or obvious cause, organs removed with apparent precision, a phenomenon that has popped up every handful of years around here since the 70s. The strangest thing about them is that the normal scavenging wildlife avoids the carcasses. Not interesting alien evidence? What about the videos

leaked from the pentagon and then confirmed of flying craft that appear to operate in ways we can't explain?

I find it difficult to accept that adult aliens intelligent enough to cross the vast interstellar voids would come here to do any of the silly things that ufologists accuse them of. Thus, I fall back on the Douglas Adams hypothesis:

Every Friday night, alien teenagers around the galaxy borrow their parent's spacecraft and warp on over to Earth to play pranks. Some like abducting the locals and spooking them with portentous messages or by having sex with them. Some enjoy barbecuing animals with lasers or making crop circles with tractor beams. The techie nerds amongst the alien kids often delight in using holographic camouflage to make their ships look weird and cool, or in hot wiring their warp drives to break the lightspeed limit and the principles of inertia.

Only an alien artifact or an alien biological form would convince me that aliens physically visit Earth.

237) How about Timothy Leary's idea that he was being sent messages or tuning into messages from another star system?

That sounds like an excellent magical technique so long as you check the messages for sense and reason and sanity. The experimental belief that your own subconscious, the collective unconscious, extraterrestrial intelligence, or some future state of yourself might know something that you don't and might be willing to tell you, can yield results. Many of those who had an inspired idea speak of it as seeming to come from 'outside' of their conscious mind. Magicians deliberately go looking 'outside' rather than waiting for something to fall in.

238) How worried about space germs should we be in exploring space?

All terrestrial lifeforms have a lot of biochemistry in common. We do not know if completely different biochemistries are possible or if only carbon-based organisms can work, or if only DNA and RNA molecules can offer enough complexity to allow life to develop. Xenobionts with completely different biochemistries to the terrestrial type should pose no disease threat at all, if they use a completely different chemistry to ours then it cannot hijack ours. If we ever land on a planet where we find something that we can eat, then we need to exercise great caution.

239) Do you think consciousness is evolving?

If we could somehow swap a modern baby with a baby from forty thousand years ago the one that came here would probably grow up and thrive (although it might have problems with resistance to modern diseases). It might well outshine most of us because people in those times had to have the ability to master a far wider range of skills than most people do today. I think they were on average tougher and smarter than us domesticated people. The baby that went back might have difficulties. This may seem an odd idea, but domesticated cats seem to have lost about a third of their brain weight and much of their ability to survive in the wild.

Consciousness always appears as an activity and it always has a subject. Today many people have more time to spend on consciousness of abstract ideas, but those ancient Greeks who had the leisure created philosophies and introspections that we have not surpassed. The Roman Philosopher-Emperor Marcus Aurelius wrote Meditations that display a depth of self-awareness that many still regard as exemplary.

We have consciousness and awareness of many phenomena that historical peoples knew nothing about, but we have lost consciousness of many of the phenomena in their world. The contents of consciousness have changed. Whether or not we have greater or lesser self-awareness remains debatable.

240) I feel like Leary's eight circuit model of consciousness was aimed at evolving self-awareness in particular, do you think it's theoretical possible?

The eight-circuit model developed by Leary and expanded upon by Wilson appears as a basically transcendental and somewhat moralistic scheme. It certainly does not map well onto what little we know about neurophysiology. Like most schemes for classifying what the mind can or should do, it looks superficially convincing, but only because the mind will tend to provide some confirmation of any scheme projected upon it – hence we have such a plethora of schools of thought in psychology. Psychological theories seem to work rather like religions, they invite you to internalize them for their apparent descriptive power, but they have very little predictive power concerning those who have not internalized them.

A few people have attempted to draw parallels between the Leary-Wilson eight-circuit model and the eight rayed star of chaos but they miss the point. The mind-star has an unlimited number of rays, humans can develop the most extraordinary range of ideas and behaviors. The eight rays of the chaostar just show eight of the big and important things you can do with your mind and your life and remind you not to neglect any of them.

241) Does evolving self-awareness come up in any white magic traditions?

'White Magic' tends to mean little more than 'not doing anything criminal or immoral'. Arguably, self-awareness implies at least an acknowledgement of one's capacity for evil. Those who refuse to acknowledge this often end up doing terrible things for 'good' reasons.

242) Do you think it's likely there are completely separate biochemical life forms?

Anaerobic life developed first on Earth because plants had not yet created free oxygen. That life still sculks around in the shadows in the form or organisms that cause tetanus and gangrene for example. Yet it shares enough biochemistry with us to enable it to attack us. No known chemical element except carbon seems to have a rich enough chemistry to support life, however complicated carbon-based molecules other than DNA or RNA and our types of proteins could perhaps serve.

243) On the other hand, it seems like even setting foot on a planet with other carbon-based life would almost certainly make you sick in some horrible way?

Probably. Keep the spacesuit on.

Look what happened to the Martians in H. G. Well's 'War of the Worlds'.

244) There are faculties of consciousness like our ability to multitask that seem dependent on physical structures of the brain since they decline with age. Do you think all the multitasking kids do these days could strengthen that ability or is that maybe going to lead to our getting dumber like a house cat?

I had what you might call a systematic school education, history from the stone age to the present. Geography of each continent in turn. English from Chaucer and Shakespeare to Orwell. Science from atomic theory to celestial mechanics. Math from Euclid to Calculus. Woodworking from straightening a stave to building a boat. Plus, the school and the scouts went camping, proper camping with axes and open fires.

It sometimes worries me that kids today get a very patchy education full of politicized topics and they spend too much time on computer games and social media to develop an extended skill set and knowledge base. There seems a crazy idea around today that there is no point in remembering anything that you can look up on the internet.

245) How exactly do you mean transcendental?

I mean transcendental in the conventional sense - immaterial phenomena which we can neither confirm not falsify.

246) What are the main ways it doesn't map well with neuropsychology?

The first four so called 'circuits' deal with observable human behavior, but a fair bit of that looks like learned behavior not innate hardware. The second four read like a transcendental wish list compiled by mystics who have vastly overestimated the effects of drugs and mistaken confused euphoria for enlightenment.

247) What do you think about Elon Musk's schemes to network the human mind?

The idea of implanting chips or electrodes to which the brain cells could pass signals to activate prosthetic and other devices could prove a boon to some disabled people. However as with most technologies we will soon think up nefarious uses as well. Fighter aircraft move so fast now that pilots barely have time to move their fingers to the weapon controls. Think and fire might seem a good idea, but I wouldn't want to fly wingman to a comrade who I had just had a bad argument with.

Humans have high neuroplasticity; we can repurpose parts of our brains and we can develop new connections within it as skills develop. Thus, it may prove possible to implant chips that we can use to develop quite complex outputs to make real-time voice synthesis devices work or perhaps to throw visualized images up on a screen.

Bypassing the senses and inputting data directly into the brain does however seem a long way off and it may long remain so. We do not understand much about how the brain encodes memory, but it does seem to involve micro restructurings of the neural architecture. If the micro-scale 'wiring' of the brain stores our memory then we would have to understand each individual's connectome, or neural map in detail before trying to modify it. Thus, I cannot suspend my disbelief in sci-fi stories where people have their 'minds' downloaded into the empty brains of clones. To do this you would have to first 'read' the donors connectome and then somehow completely restructure all the billions of neuronal pathways in the recipient brain to match.

248) Well neither of them was super knowledgeable about computers, Wilson acknowledged later that circuit was not the best metaphor but they definitely thought that the first 4 circuits were learned behavior. The last 4 are certainly fuzzy through their evolution and Leary was certainly on a euphoria trip but by the end I think Wilson did a good job of distinguishing euphoria as a step that must be passed on the way to enlightenment and even distinguished between transcending your learned programming and genetic programming. Which I think Leary got from the idea of past life karma. While it is moralistic in taking a strong stance on the power of love and the desire for harmony it still feels like a pretty solid open model for developing self-awareness. I also think Wilson felt that drugs were a tool but no substitute for the hard work of meditation, yoga and magic. It's not hard science but as a mystical model of human experience and possible potential is it all that lacking?

The four 'higher' circuits of the Leary-Wilson 8-circuit brain

theory provide an inspirational schematic for possible future developments of the human mind. They encourage us to enjoy life (5), to know ourselves (6), to know our own biogenetics (7), and to understand the entire universe (8).

Call me a curmudgeonly old cynic, but I take a rather more pessimistic view of human nature and incline towards a more Lovecraftian view of the cosmos.

Much of what constitutes our enjoyment of life involves the detriment of others, the exploitation of other species, and the degradation of the environment. We appear as the apex of a parasitical, bloody, and genocidal process of evolution, and our own 'civilisations' have evolved by exactly the same mechanisms. Those who know themselves and know what they want from life usually seem more objectionable than those who just bumble along with what they have. The ability to manipulate our own biogenetics will open many terrible cans of worms. The cosmos remains indifferent to us, it contains no gods with an interest in saving us from ourselves or natural phenomena. Nearly all parts of the cosmos seem actively hostile to life as it consists mainly of radiation blasted freezing vacuum with a thin smattering of arid rocks and white-hot stars.

Enlightenment about any topic usually encourages us to exploit it, and if possible, to weaponize it. Progress seems merely the mechanism by which we exchange one problem for many more.

Nevertheless, all this does make life interesting, sometimes too interesting, but evolution does not seem to have designed us for contentment with anything for long.

We probably stand on the brink of awesome discoveries about what we can do with our own biology, with mind control, and with the manipulation of space and time and matter and energy. We shall undoubtedly do some grim things with these powers. If we survive as a species, I suspect we shall come to resemble something closer to Lovecraft's Elder Gods than to a race of Buddhist Bodhisattvas.

Yet we shall probably acquire the option of modifying human nature by genetic tinkering. The consequences of doing so seem presently incalculable. Would we choose to make it more cooperative or more competitive?

Lifespan will probably prove relatively easy to modify. We do not die of entropy; we age and die because we have evolved genetic programs that kill us. Each species evolves a lifespan that represents a tradeoff between living long enough to reproduce but not living long enough to inhibit the evolution of the species. Some big reptiles and deep-sea fish live for centuries. We could soon probably make that millennia for ourselves.

249) It seems like there's a lot of weird ass middle ground between uploading data right to the brain and those tongue stimulated sight set ups, I mean the brain can learn to make sense out of all sorts of stimuli, right? Couldn't it eventually learn to build an image out binary input or would that require a ton of storage or something?

I can imagine a bodysuit that could deliver tactile stimuli to any part of the body thus turning the entire body into a large screen onto which we could project information. Apparently, blind people can learn to interpret visual information from a camera on their chest or head made into tactile stimuli projected on to parts of their skin. Braille readers can learn to read by running their fingers over surfaces with little bumps representing letters on them. It might take a lot of training but potentially we could learn to read a report or watch a film projected in tactile form on to our backs or chests.

250) Do you think that could enable us to communicate with each other on something close to a psychic level using pictures and feelings rather than words?

Artists have always sought to get inside our heads with their creations. Various forms of virtual reality may offer many more possibilities for them. CGI, computer generated imagery, has spoiled many cinematic experiences for me when it has become too obvious or too ridiculous or a substitute for quality drama. Yet as it gets better technically, 'deep fake' technology will open the doors of gullibility and paranoid skepticism.

251) The way I interpreted the first four circuits is with the design concept of affordability, a certain set up affords use in a particular way, hence distinctions between the "reptilian" and "mammalian" brains, these different configurations afford opportunities for different types of learned awareness. Is that totally off base or maybe a limiting way to look at it?

The human brain certainly shows signs of having grown to its present design by evolutionary accretion, it still contains parts which look remarkably similar in form and function to parts found in fish, amphibians, reptiles, and monkeys. The two apes closest to us, the chimpanzees, and the bonobos, both exhibit social behavior that closely resembles ours, with the chimpanzees tending to display far more of our 'worse' aggressive and violent behavior than the bonobos.

Eliphas Levi portrayed his vision of 'god' or 'a human writ large' as Baphomet, an androgynous humanoid with additional features corresponding to other mammals and to reptiles.

Austin Spare mentions 'Atavistic Resurgence', the liberation of the powers of the animal forms we have evolved through and still carry traces of in our brains and bodies. Unfortunately, he did not elaborate in comprehensible form upon his techniques.

Running through all religious and mystical systems we see a tension between transcendence of our animal natures and transcendence through our animal natures. Levi abandoned a vocation in the catholic priesthood because he fell in love, and he went on to create in Baphomet a godform that looks remarkably like the christian devil. Both Austin Spare and Aleister Crowley tried incorporating their sexuality into their mysticism and magic. Conventions and rules about sex and diet characterize all religions, cults, and cultures.

Our large cerebral cortexes seem to have developed on top of the basic mammalian brain in tandem with our development of language and complex socialization and our use of tools. These three activities demand a massive information processing capacity that took several million years to develop. Once we had acquired such a capacity; philosophy, higher mathematics, all the arts and sciences, and rockets to the moon all appeared as side effects in a mere eyeblink of evolutionary time.

252) What about dropping crime rates, a still growing consensus that war is not good and the popular attempts to address social inequality? You don't think that these are signs we are moving away from some of our nastier qualities?

Crime statistics depend on the definition, the reporting, and the recording of crime. In the UK, the authorities have currently prioritized the suppression of violence and any expression of discrimination or 'hate crime'. Consequently, the statistics for such crimes have gone through the roof. The actual levels of violence have probably remained roughly constant although in recent years the incidences of teenagers stabbing each other seems to have risen markedly. Overall, it seems about as dangerous to walk around at night as ever, although the omnipresent CCTV has made urban centers a bit safer.

In the UK, the authorities now assign low priority to property crime and fraud and these have become daily facts of life that few people even bother to report anymore. Everyone here gets bombarded with fraudulent telephone calls and scam electronic messages and anything left unguarded rapidly gets vandalized or stolen. Trust and social cohesion certainly seem to have declined but we have no metric for that.

Our attitudes to war remain as ambiguous as ever. Violence remains a staple of the entertainment industry. Global expenditure on 'Defense' seems to rise continually but old-fashioned style war had tended to decline in popularity because it has become increasingly uneconomic. At present it remains exceptionally difficult to turn a profit from invading someone else's country because current weapons technology heavily favors both conventional and irregular defense. Let us hope it stays that way. If offensive technologies become available and profitable to use it won't take much patriotic propaganda and jingoism to get us using them.

Social inequality remains as popular as ever, most people still strive to improve their lot relative to others. The new clique of social justice warriors comes almost entirely from the ranks of the now downwardly mobile lower middle classes. Their prospects have become squeezed, and they seek a new power base.

Interview 8

253) Do you think that meditation or anything like that can actually make us more caring and self-aware?

It all depends upon what you meditate about. Zen Buddhism underlaid the militaristic Japanese Samurai tradition of Bushido. No-Mind makes it easier to put a sword through a victim or your own guts.

Meditation seems to mean one of two things, either thinking about something in particular, or striving for an inner void where you do not think of anything.

Meditating about compassion can obviously increase your propensity to exhibit it and achieving no-mind can perhaps increase it because compassion always has costs involving some degree of ignoring your own interests. Thus, self-awareness can decrease compassion.

The Tibetan Buddhists preach endlessly about compassion. Compassion came at a high price on the Tibetan Plateau, particularly in winter. Feed and shelter a starving and frozen neighbor or traveler in those parts and you risk severely compromising your own survival. When you hear a virtue continually exonerated, expect a shortage of it. Avoid cultures that advertise a 'Merciful' deity!

254) What do you think about GDP or its alternatives like national happiness indexes?

Various methods exist for calculating Gross Domestic Product, most of them count wealth and illth as one, and just give the overall economic churn. Here illth implies the circulation of money and work which contributes negatively to general wellbeing, for example pollution creation, drug abuse, and gambling. Such expenditures expenditure all push up GDP. Waste and obsolescence of course do great things for

GDP totals. In the end GDP merely measures how busy we have become and how much stuff we use.

Tragically we have constructed fiduciary monetary systems financed entirely by debt and predicated upon economic growth and become hopelessly and perhaps terminally addicted to them. We have evolved a system that has to expand continually to prevent itself from imploding.*

This system exists because of our belief in it. Whatever else anyone believes in, almost everyone believes in fiduciary money and the necessity of economic growth.

Yet underneath this belief lurks the doubt and the rational certainty that we cannot sustain this.

It will take several Economic Messiahs and/or multiple manmade catastrophes to provoke a paradigm change.

Gross National Happiness sounds like a splendid ideal, but who will define the happiness to which we will become subjected? Some people consider that Bhutan has set an example with GNH.

The Bhutanese government simplified the problem for itself by ethnically cleansing the country of Hindus who didn't accept their Buddhist definition of happiness.

(* I have devoted an enormous amount of time to demonstrating that the universe does not need to expand to prevent itself from imploding, but that seems a comparatively simple problem compared to convincing humanity to adopt a rotating steady state economy.)

255) Wait, what's a rotating steady state economy?

My alternative to an expanding universe has a form of rotation which stabilizes it against implosion and allows it to remain a steady size. This model, called Hypersphere Cosmology, has no time limits so the whole universe continually recycles itself, stars and galaxies and black holes all come and go, everything gets recycled. The recently discovered vast gas and dust halos around galaxies probably function as the recycling yards of those galaxies.

By analogy, a Rotating Steady State Economy would stay the same size and everything in it would have to become recycled and driven by renewable energy sources only. As this world acts as a thermodynamically open system rather than as the closed system of the entire universe, we must interpret renewable energy as basically any source that comes ultimately from the sun. We should not do anything that we cannot undo using available renewable energy. This means no more burning of fossil fuels or radionucleotides and using only fully recyclable building materials and products.

The principles seem simple, persuading humanity to adopt them will prove tough.

Adopting renewable energy itself as a currency might help a lot. This would allow forms of growth based on the growth of the real capacity to generate it, rather than on the capacity to destroy more irreplaceable resources.

256) Would that have any inherent effect on inequality? Or black markets?

Liberty! Equality! Fraternity! The old French Revolutionary slogan sums up the political paradox rather concisely. Liberty and Equality contradict each other. Give people Liberty and they rapidly become Unequal.

Fraternity rarely has much effect beyond Dunbar's Number, we seem to have evolved the capacity to empathize with about 150 people at maximum.

Didn't some ancient Greek note that the roots of war lay in the tendency to favor our own children over the children of others?

Many seem keen on the idea of equality with those who have more, very few seem keen on equality with those who have less. I always remain highly suspicious of those who preach equality.

Wealth seems to act rather like heat in thermodynamics, it can only do useful work if unevenly distributed.

However, the most successful systems have some mechanisms to prevent the thermodynamic anomaly of heat flowing from cold to hot. Wealth should not flow automatically towards wealth. It should take work to create and maintain wealth. Most capitalist economies recognize this and have some form of progressive taxation and death duties.

Systems of enforced equality have not worked well. Communist regimes have always had to resort to motivating people with bayonets, and those in charge of them have invariably had far more privileges than those subject to them.

Any system involving humans will involve screw-ups and unforeseen consequences. However, we do know that any law will create a criminal opportunity and that any rule will create some sort of a black market.

257) Well, it does seem like there's a point at which income inequality gets so significant that productivity declines?

I would guess that a graph of equality against productivity would show a Bell Curve, at the extreme inequality level productivity breaks down and sabotage and revolution begin. At the extreme equality level nobody gives a damn because effort becomes similarly disconnected from reward. As they used to say in the old eastern bloc countries, 'we pretend to work, and they pretend to pay us'. At the extremes you can only make people act productively by non-economic forms of coercion, usually naked force. Yet the Soviet system did manage an astonishing productive effort by also invoking the fraternity of patriotism during WW2.

258) What would be the first moves in that direction?

To achieve a Rotating Steady State Economy (RSSE), we would either have to achieve a global consensus and an enforced moratorium on the use of fossil fuels, OR, we will have to develop renewable energy technology to the point where it easily outperforms fossil fuel technology in terms of effort and cost.

259) Investment in renewable energy and creating an actual recycling system?

RSSE requires not only investment but research. A lot of the recycling currently performed seems counterproductive and merely aims to reduce landfill at an extremely high energy cost. We need to address the problems at source and stop manufacturing single use glass items and to stop manufacturing biologically inert plastics entirely.

260) How would a currency based on renewable energy work?

The demons always lie in the details. Nobody 'understands' quantum theory, and nobody 'understands' economics either, for similar reasons – both fields deal with seemingly stupefyingly complex phenomena that also remain subject to observer and experimenter effects. Economic theories tend to distort the very systems they purport to describe.

Currently, the value of any currency effectively depends on the ability of its issuing nation to metabolize and destroy non-renewable energy sources. This has increasingly catastrophic consequences.

The experiment of adopting renewable energy itself, say in kilowatt-hours, as the basis of currency might have some beneficial effects, particularly if it led to the pricing of goods and services in terms of their true energy cost. In principle, the economy could then only grow by increasing its renewable energy generating capacity. A 'banknote' for say a megawatt-hour would entitle its bearer to draw precisely that amount of energy from the 'bank' or reinvest it with the bank to develop more generating capacity or exchange it for goods or services of that energy cost.

Do we want a fundamentally destructive entropy increasing economy or a fundamentally constructive negentropy economy?

261) At this point what do you think the most promising renewable technologies are?

Hydroelectric, wind, tidal, and solar power all work well and need much more investment, plus we need to invest in off-peak power storage for the last three intermittent sources of power. Pumping water uphill, air compression or air liquefaction all show promise. Electrical storage batteries will only serve us well if we can up their efficiency and reduce their use of expensive rare metals.

262) Where should renewable research focus?

Solar-voltaic installations have the limitation that they tend to produce electrical power only in strong sunlight and often only during times of reduced demand. I keep coming across references to research into solar-catalytic processes and various solar-photosynthetic processes to create storable fuels like hydrogen, methane, methanol or ethanol or other hydrocarbons using only moderate levels of sunlight. We could distribute and use such fuels with only minor modifications to existing infrastructure and machinery. We know these technologies work on the laboratory scale and there seem few barriers to scaling them up except vested interests and a lack of imagination, investment, and foresight.

These technologies deserve far more research and investment than 'hot' nuclear fusion which we have already exhaustively found almost impossible to scale down from a reactor the size of a star.

'Cold' forms of nuclear fusion just might prove possible although we have few clues about how to even try, but we should keep theorizing and tinkering with fundamental particle physics.

263) So, would the RSSE help keep the dispersion of wealth in the productive part of the bell curve?

The adoption of RSSE would imply that a recognition of the finite resource capacity of this planet had occurred. Resource depleting economics would have to become prevented by law and force.

Once we recognize that economics has to become an almost zero-sum game with only modest growth allowed through increasing renewable energy capacity, then everything changes:

Debt fueled growth would come to an end. The debts we pile up today essentially consist of spending the future – we currently sell off the sustainable future of our habitat.

Productivity would have to decline, we have far too much of it anyway and most of it creates waste rather than permanent assets. The oceans fill with plastics, the skies fill with pollution, most of what we build and make now has built in obsolescence and poor recyclability.

People will have to get used to less work and more leisure. To finding fulfilment in social activities and personal development rather than in material status.

Zero-sum games have usually set the scene for war. If you cannot make more it becomes tempting to take what others have. I take some comfort from technical developments that have increasingly made war a negative-sum game. I do entertain the hope that societies can develop mechanisms to turn the accumulation of excessive wealth into a negative-sum game rather than a lauded achievement. Wealth should take exponentially increasing effort to maintain, rather than automatically attract more to itself.

264) Are you into Bucky Fuller's ideas about a shared global electric grid?

The idealism seems laudable, but it all seems predicated upon electricity becoming 'almost free' and an extraordinary international consensus on what we can and cannot do with it.

Presently all humans live under a spectrum of regimes that runs from those where dictators and their subjects live in mutual fear of each other to those which select their leaders on their ability or promises to make more of everything available.

265) I think a lot of people feel like we're waiting for that one, or maybe a few, technological innovations that will make one or more renewable energy sources economically convenient. Is that a problematic way of looking at it? Or if not, what do you think that innovation will have to be?

Renewable energy technology alone might just reduce climate catastrophe by mid twenty first century, but no imaginable technical fixes seem likely to ameliorate the other resource catastrophes we will face by the end of the century, if we then have a world population of twelve billion who all want to consume at even the present level of Europeans or Americans. Depletion of freshwater reserves and soil erosion appear increasingly serious.

Technology has got us into a crisis situation, it will take more than just technology to get us out of it, it will take politics and a philosophical rethink about what we can and should do as custodians of this planet.

266) In what ways do you think the average European or American lifestyle has to change? Maybe what area do you think we waste most in? Transportation or just buying cheap crap we don't need or wasteful food habits?

All these things obviously; but underlying all these behaviors seems the modern obsession with measuring ourselves relative to others. The pursuit of status and pleasure never brings any lasting satisfaction.

Culture and advertising have such an extraordinary power to define what pleasure should consist of that many unimaginative people do not even get much pleasure out of their pleasures.

Depression haunts the consumer society.

I would welcome a cultural shift towards measuring ourselves relative to ourselves. Only the pursuit of virtue brings lasting satisfaction as some of the ancient Greeks realized.

Virtue here means the acquisition of genuine knowledge, wisdom, ability, and the higher regard (but not the fear or envy) of your fellows.

The pursuit of virtue offers a more challenging and rewarding path, but our culture currently tends to promote the easy options for commercial reasons.

267) Does consensus reality have an impact on magical probability?

Belief has a strong influence on what we can and cannot do, and on what we can and cannot perceive.

Cultural consensus has a strong influence on what we can and cannot do, and on what we can and cannot perceive.

Magicians have always striven to transcend beliefs and consensuses so that they can do and perceive more.

268) Do magical probabilities change at all during times of cultural and social uncertainty and upheaval?

Many periods of cultural and social upheaval have seen an upsurge in magical thinking. This arises because we find it more effective to deal with a cultural or social phenomena as some sort of a 'force' or 'entity' which has a mind of its own, which of course it does, because it has hijacked some of our minds. Political and moral fashions come and go over the decades and centuries, and after they have gone, we wonder how we ever became possessed by them and all the now seemingly strange attitudes they brought with them.

269) Do coincidences or synchronicities have any meaning or use in chaos magic?

Hardcore magicians consider that everything happens by Magic and that Science just deals with those events which occur fairly reliably and repeatedly. Magicians tend to work at the other end of the spectrum where events become less reliable and trickier to engineer. Thus, all apparent cause and effect relationships consist of reliable coincidences, and all strange coincidences and synchronicities suggest another form of magical connection worth investigating.

Personally, I prefer a magical rather than a strictly causal explanation of reality because any causal scheme must either imply an infinite regress or an infinite divisibility of reality or some level where events just magically happen out of themselves so to speak.

270) What do you think about Israel Regardie's advice that one should spend some time in therapy before trying magic?

We Brits tend to regard psychoanalysis as a silly self-indulgence of the American middle classes. There seems no evidence that it can cure mental illness or that it does any more for mental wellbeing than a brisk walk and meeting with friends. Austin Spare referred to Jung and Freud as junk and fraud. Both Jung and Freud gave up on trying to heal the mad and took to peddling pseudo-religious psychobabble to the merely unhappy wealthy.

Regardie's advice came loaded with self-interest. He used to peddle such therapy.

Our understanding of human psychology has not advanced since the time of the ancient Greeks. A reading of the classics and the philosophers can tell us all that we know about ourselves. Modern psychology consists of ad hoc hypotheses dressed up in mock scientific language and backed up by ludicrously unscientific research.

Nevertheless, Regardie's advice does point towards something important – people who cannot even handle ordinary reality should stay away from magic. Never give a sword to someone who cannot even dance.

I have seen many people with screwed up lives take to magic in the expectation that it will solve their problems. It rarely does, it usually just makes things worse by amplifying the faults.

Chaos Magic represents a shortening of the path in The Way of Magic, it has made the path steeper and more direct and much easier to fall off. Beware - it has a high casualty rate.

Only those who can function well in ordinary ways and who know what they want to achieve should employ magic to enhance their chances.

Always beware of those 'mages' who peddle magic as psychotherapy, this provides a sure sign that they couldn't actually get it to work for themselves or show anyone else how to.

271) How would you define magical thinking distinct from the practice of magic?

Magical Thinking means perceiving connections between phenomena that according to conventional thinking have no connection. It also means trying to create connections between phenomena that seemingly have none, and this seems so unthinkable to conventional thinkers that they seldom acknowledge it.

Practical Magic means trying to use those connections.

Conventional Thinking usually refers to Magical Thinking in a derisory fashion, often describing it as superstition or delusion, or even as mental illness. Yet just about all of science developed out of people investigating connections that other people did not realize existed. A great deal of what goes on in the human realms of politics, fashion, religion, advertising, morality, and personal identity, depends on the active creation of associations and connections between phenomena previously regarded as unconnected.

Magical Thinking means Apophenia – the perceiving or imposing of hidden or occult connections, and in practice it dominates our thinking because it both establishes conventional thinking in the first place and then it seeks to overturn and upgrade it.

I worship at the feet of Apophenia who I personalize as a Greek style goddess, the Muse of Imagination. As a magical thinker, if I want inspiration, I simply personify it and invoke it. She provokes some extraordinary thoughts and inspirations, some of them a bit too extraordinary, but these may come from her crazy sister Pareidolia.

272) So, I was sort of thinking chaos magic would insulate one from the pitfalls of obsession but you're saying it's more dangerous?

In Chaos Magic there seems less danger of obsession with the foci of Invocation because of the treatment of belief as a tool rather than as an end in itself. However, treating belief in this way can lead to the acquisition of unrealistic beliefs about oneself, particularly after beginner's luck with early spells. Arrogance seems a particular problem, I have seen it lead to failure and disaster many times.

273) How does your definition of archetypes differ from Jung?

Jungian archetypes have always remained rather loosely and imprecisely defined. They seem to imply everything from Platonic 'Forms' to Biological Instincts, to Mythological and Cultural and Social Constructs. The underlying thinking here seems to say that humans imagine and invent gods and goddesses to represent existing human social roles and that we can look to these forms as role models for our thoughts and feelings and behaviors, and perhaps also for some cautionary tales and folk wisdom. Today of course celebrities play a major role in supplying us with archetypes almost as imaginary as those from mythology.

Chaos magic took this a step further and challenged people to invoke archetypes from fantasy sources and science fiction as well. The underlying thinking here says that as we now live in a world where identities and roles evolve fast, we need to imagine new archetypes for them. The futuristic Elder Gods of H. P. Lovecraft's Cthulhu Mythos may serve as well as the Olympian Deities of the Classical Greek Mythos.

274) So how much magic would you say you do on a daily basis?

Since retiring from the frenzy of commerce into the more sedate business of property management, and since retiring from the management of magical orders, I have devoted most of my magical efforts to the preservation of what I have acquired and the quest for more knowledge for its own sake.

Preservation begins with my own body and mind. Daily I perform strength maintaining exercises combined with visualizations to direct 'chi' or 'prana' or 'life force' or 'magical intent' or whatever else you want to call it, to all parts of my organism. A lot of would-be magicians get things the wrong way around and fail to look after their prime instruments – their own minds and bodies. Usually more than once a day I evoke by visualization my main servitor to perform protective functions for those phenomena and people dear to me.

In a quest for knowledge, I invoke into myself the inspirations of Apophenia, and lately those of some of the Elder Gods also. These invocations may range from brief visualizations and incantations to more elaborate rituals with instruments other paraphernalia, depending on the difficulties I find myself wrestling with.

I do not count the several hours per day I now spend on magical correspondence and physical/metaphysical research as 'doing magic', yet they do form part of my continuing magical quest.

Having acquired an abundance of material possessions and personal relationships I rarely enchant for enhancements to my own circumstances anymore. The days of spells to back up every business deal have long passed, yet I still occasionally throw in the odd enchantment to influence events in the wider world and quite frequently one to stop it raining during my long walks. (A perennial problem in these Atlantic storm lashed isles.)

275) Is your current level of daily magic less than during the frenzy of commerce years or just different?

During my 'frenzy of commerce' years I also raised a family and developed a lot of practical material for the IOT and then for Arcanorium College.

Now in retirement I have time to philosophize more deeply about all the ideas that I either roughly appropriated for use, or which came out of what we did and experienced, or which have developed in science in recent decades. The seventeenth century natural magician Thomas Vaughan quipped 'A Witch is a rebel in physics.' I seem to spend increasing amount of the likely twenty years that I have left, trying to work out the remaining secrets of the universe by the methods of Natural Philosophy

– and this means using inputs from both science and magic.

I seem busier than ever, my wife considers me somewhat obsessed, particularly when I do not free up enough time to assist her with her obsessions in horticulture.

276) Do you have a personal definition of obsession you use to gage maybe how reasonable or healthy a level of interest is?

I regard the ability to concentrate on something deeply for long periods a great virtue, so long as you can turn it off and laugh at it and do something else or do something completely contrary to it at any time.

The Thelemic idea of having a 'True Will' or 'Holy Guardian Angel' has always seemed to me one of the worst possible ideas used in magic. It seems a religious rather than a magical idea, it basically consists of the idea of 'original sin' turned on its head. It implies that you have a single purpose in life and that you must discover it and dedicate your entire life to it. For most followers of Crowleyanity this means a slavish imitation of Crowley which never gives good results. It merely turns the magician into the demon Choronzon and leads to an early demise.

277) So where is the line between magical thinking and insanity?

Any form of thinking taken too far can result in crazy behavior. Kurt Godel, one of the 20th century's greatest logicians could barely function in the real world and died of self-neglect.

Even simply taking the idea of cause and effect too far can drive people mad. It can disable the sense of free will and self and remove all motivation or sense of responsibility.

Magical thinking can go too far, particularly if you take the infamous and aptly named Oath of the Abyss; "I will interpret every phenomenon as a particular dealing of God with my soul." That opens the door to paranoia and/or megalomania and delusion. Heck, some things just remain random or meaningless.

On the other hand, a little judicious use of magical thinking can prove invaluable. If something just doesn't feel right or does feel particularly right, despite the mundane evidence otherwise, then your subconscious may well know better than your conscious. Good magical thinking often comes down to getting the best out of your intuition.

278) Do you have any advice for developing or interpreting intuition?

All the techniques of Divination should improve intuition, even if they do not appear to yield immediate useful divinatory results. Any form of mind-stilling meditation should improve it. Paying more attention to dreams and daydreams, particularly the non-verbal parts of them, can also help.

However, the majority of intuitions will turn out wrong, so always try to give them an intuition check (i.e., against longer term intuitions that have proved themselves) or a sanity check to see if reality could accommodate them, before acting upon them.

279) What are your thoughts on Reich's general ideas about sexual repression on both the social and individual levels?

Dissatisfaction seems built into our sexuality for good evolutionary reasons. Neither Monogamy, Polyandry, nor Celibacy, nor Libertinism seem to bring long term satisfaction to all people of all ages, or to all cultures. Sexual moralities have their origins in the political, economic, technical, and social structures that we create, they do not tend to lead them. Yet no system seems to work to everyone's satisfaction, so we end up with priests and prostitutes, rule breakers and rule makers, and all manner of prophets, gurus, charlatans, hucksters, and lately therapists, trying to make a living from selling us our own wallets. Having big brains does not lead to any clarity, it just makes it all more complicated.

Rather than add to the eternal debate about what we should or should not believe or do, or what constitutes liberation or repression, I will simply summarize the two main points about sex magic.

Firstly, we have the simple equation between orgasm and gnosis. The height of sexual excitement can provide a useful altered state for the casting of spells, or the reprogramming of beliefs, or the reception of unusual perceptions and other magical maneuvers.

Secondly, sex and procreation underlie a huge amount of our motivation. So much of what we do has its roots in trying to prove our reproductive fitness and improve our attractiveness. We have come up with some extraordinary ways of attempting this. Wealth, power, fame, genius, philanthropy, fashion, and moral or physical heroism all act as imagined aphrodisiacs that people pursue to enhance their imagined worth and various sublimations of the deeply biological urge to achieve immortality of sorts by procreation.

Thus, magicians often put enormous value on the idea of 'The Muse' – real OR imagined partners to whom they dedicate their work and achievements in the hope that this will inspire them further. This can have spectacular but incendiary and short-lived effects if the real AND imagined partner do not comfortably coincide.

280) So, if talk therapy and orgies aren't the answer how should societies help the troubled or insane?

Talk therapy and orgies may well have beneficial effects on some of those whose sexual inclinations slightly mismatch social expectations, but no form of psychiatry seems reliably effective with the seriously insane. There seems little evidence that psychiatry makes much difference to recovery times, so it mainly just acts through chemical suppression or physical containment.

281) It seems like we have a mass shooting every week now in the land of the free, is that just an issue of American individualist philosophy and access to guns?

Obviously, it's the guns. We have murderers and attempted mass murderers here, but they rarely get far with mass murder using clubs and knives. Keep the world safe for hand-to-hand combat, it takes more skill and far more determination. Guns make killing stupidly easy.

The American arguments for free possession of guns seem crazy to us Brits. The argument that guns protect individuals from an overmighty state implies that individuals should have their own anti-tank and anti-aircraft missiles, and maybe even their own nukes.

282) Well sure but we had plenty of guns five years ago and we didn't have a mass shooting a week, what's the increase due to?

Demonic possession.

From this side of the Atlantic, America seems a land of extremes where opinions and identities have become very polarized, and debate has become poisonous and ridden with culture war. The new electronic media have worsened the situation as anyone can now find an echo chamber online that will feedback and amplify any point of view without criticism.

I would call all this seriously bad magical thinking. In magic we have the dictum 'Invoke Often' and this also means widely, we should try invoking all sorts of different things to gain a broad-spectrum perspective. If we continually seek only confirmation and reinforcement of our own prejudices and our beliefs about ourselves, we will turn them into demons.

Demonic possession has one defining characteristic – it makes you regard those that disagree with you or oppose you as demons.

283) Tantra seems to me like more complicated sexual mysticism than like Greco-Roman mystery cults, tantra sort of reminds me of Reichs conception of sexual energy? Are there some basic variations on sexual magic traditions?

Hinduism and Buddhism have evolved a hugely complicated array of ideas and practices and sects over the millennia. Some of these focus on the use of the body and the senses to achieve mystical experiences and sometimes magical effects. Some western commentators took a particular interest in the sects that used sexual ideas and imagery and sexual practices and/or those which used antinomian practices which deliberately broke social mores in the pursuit of enlightenment, liberation, and/or magical abilities. These they classified as 'Tantric', despite that few Easterners regard all the ideas and practices as forming a coherent whole across Hindu and Buddhist cultures. In short, there seems an enormous variety of ideas and practices out east that some western commentators classify as Tantric to some degree, it can include Yoga for example.

The prurient western curiosity about sexual Tantra owes a lot to centuries of denial and mortification of the physical body under Christianity, but on the other hand a lot of modern and educated Asians take a dim view of Tantra, particularly after colonialization by Victorian era Europeans. Because it involves humans and religions, Tantra provides endless opportunities for charlatans and con artists to make a living and have fun. I rapidly grew weary of the ridiculous antics of so-called holy men during two years in India.

The tantric idea of Kundalini does seem to overlap with that of Reich's Orgone-Energy in the sense that both remain vaguely defined but associated with ideas of vitality, motivation, and sexual enthusiasm.

The Greco-Roman mystery cults do not seem to have involved much sexual behavior, perhaps because Hellenic culture in general seemed to place so few restrictions on it, until of course Christianity came along.

From what I have heard of sexual tantra from practitioners and scholars, it uses the same two effects that we recognize in modern western magic – the gnosis of arousal, and the inspiring effects of devotion to the Muse. The traditional western esoteric cannon seems to have made little use of one important 'tantric' technique – that of breath control, but in the last century or so it has adopted pranayama and related techniques.

Overtly sexual magical and mystical practices seem rare in the texts of traditional western esoterics and thus many people regard them as eastern inspired imports. However, in Christendom, such practices were frequently attributed to the real or imagined enemies of orthodoxy such as heretics or witches. Many of the surviving texts on alchemy do appear to contain a lot of coded references to sexual symbolism and practices. Many 'secret societies' within Christendom became accused or suspected of unorthodox uses of sexuality, perhaps we shall never know what really went on.

284) What does breath control add to the mix?

In the west, we had little acknowledgement of the extraordinarily strong relationship between rate and depth of breathing and thought, emotion and mental states. What little acknowledgement we had seemed mainly related to song and chanting and the general idea of taking a few deep breaths to calm down and hyperventilation as a sign of agitation. In the east, the regulation of breathing to control thought and emotion developed into an extensive series of practices for evoking a wide range of mental states from the excitement of kundalini to the quiescence of divinatory trance. In preparing for a chaos magic ritual, we always consider the breathing pattern that will best compliment the rite, for that does half the work of creating the appropriate mental state in itself.

Interview 9

285) How would one perform an exorcism on a country?

You can try eliminating demonic possession and its proclivity for demonizing its opponents by destroying Symbols, Institutions, and Leaders. It worked well enough in the de-Nazification of post WW2 Germany for example. Conquerors and revolutionaries have always used such tactics to change a society's obsessions to their own. However, this does not work if you have a cultural civil war in which no side can achieve complete victory.

A society that has merely acquired demonic levels of polarization and culture war has suffered a serious failure of sense of humor, and for this you need Laughter Banishing.

Self-deprecatory humor and humorous rather than vicious mockery of opponents can work wonders, and there seems a tragic loss of both in certain countries at present. Many can no longer laugh at themselves, many subjects for humor have become illicit, everyone takes offense far too easily.

Those who cannot laugh at themselves become furious when others do, and polarization becomes worse. I would like to see the entertainment industry featuring representatives of all social classes and political and religious and ethnic persuasions making mild fun of themselves and their beliefs and behaviors.

I can remember a time when the media did a lot of this, it seems sadly lacking today. The media have a social duty to do more. There seem fewer social lubricants more important than humor.

286) So, what's the difference between focusing your consciousness

on humanitarian values and what you referred to as the moralistic part of the eight-circuit model?

Belief in the eight-circuit model can certainly confer feelings of moral superiority and self-righteousness, as with any religion proffering 'higher' this or 'higher' that; and suitable feelings of sin and guilt when you do not live up to it. Plus, you also get camaraderie with fellow believers, and best of all you get to pity or despise the unpersuaded. Unlike the Greek or Norse Myths, it contains no laughs.

I remain unpersuaded of the discrete separate existence of those circuits or that their supposed properties make any of them necessarily 'higher' than any of the others. The whole edifice looks like an ad hoc psycho-spiritual hierarchy cobbled together out of Freudianism, Kabala, and Science Fiction. I have seen far worse esoteric moral wish lists, but I would caution anyone about taking the drugs recommended to access some of the alleged circuits. I have met people who have done bucketloads of those substances and who seem even less enlightened than when they started.

Humanitarian action should always concentrate on the basics rather than on 'higher' considerations. Too often those motivated by higher considerations end up doing 'bad' things for what they consider 'good' reasons.

I think we would have far less need of conflict resolution if we cultivated a wider sense of humor.

287) What are the kabalistic aspects of the 8 circuits?

The eight-circuit model follows the same basic pattern as kabalistic schemes of sephiroth that developed in the early centuries AD. At the bottom you have the earthly and emotional levels and, in the middle, the psychocosm of the mental 'gifts of the gods', and at the top you have some sort of ill-defined union with cosmic consciousness or cosmic being. The Gnostics developed similar schemes in the same period of fertile interface between Paganism, Neo-Platonism, and Judaic inspired monotheism.

288) What are the funniest traditions?

The non-monotheist religions. The pagan and polytheistic religions had very human style gods who squabbled and played tricks and fornicated with each other. Laughter and humor played a part in many rituals, particularly fertility rituals. The public recounting of those mythologies almost certainly left gaps for audience laughter at some of the antics of their own gods and goddesses.

Oriental non-theistic systems like Taoism and Zen have a place for humor and the figure of the Laughing Buddha remains revered.

Nobody laughs in the mythologies of the three Abrahamic monotheisms; and laughing at them remains a blasphemous and dangerous activity in many places.

Discordianism and The Cult of the Flying Spaghetti Monster both developed as jokes masquerading as religions and/or religions masquerading as jokes, to undermine the humorless monotheism of the culture from which they sprang in rebellion. All hail Eris!

289) Do the trickster deities exist to inject humor, or do they serve another or wider purpose?

They do all sorts of things. They variously steal from the other gods, outwit them, ridicule them, cause conflict amongst them, and innovate things for them, and they do much the same for humans. They embody the principles of questioning and upsetting the status quo.

In traditional cultures the activities of trickster figures usually remain circumscribed within limits and tricksters often get punished for upsetting the established order of things. However, the trickster Loki ends up destroying all the gods in some versions of the Norse myths.

Chaos Magicians often proclaim that Nothing is True, and Everything is Permitted, or Nothing has Ultimate Truth and Anything Remains Possible, or All Truths Remain Frame Dependent and All Frames Remain Breakable.

Thus, Chaos Magicians often revel in the antinomian order-upending deities and identities such as Ouranos, Eris, Pandora, Baphomet, Apophenia, and Azathoth in the expectation of changing themselves and/or the world when the existing order seems unsatisfactory. Mercury, the traditional deity of magicians and tricksters did not seem to have a wide enough remit for everything that contemporary magicians wanted to attempt.

290) What's important to note about the relationship between magic and religion? Is it an evolutionary relationship or is there a hard line between the two?

People cook up religions when they fail at science or magic. When they fail to understand and control the world, they settle for merely controlling other people.

Understanding and controlling the world has proved a fantastically challenging and difficult task and we have spent hundreds of thousands of years working out how to make tools out of wood and stone and later metals, working out how to hunt effectively and then how to farm and to build, working out how to ameliorate the effects of disease, working out how to navigate, working out how to make machines, and so on. Achieving these things required science in the broadest sense, we had to understand how stuff worked, it also required astonishing leaps of imagination and intent that most other animals cannot manage.

For much of the time we made mistakes, we came up with wrong explanations, we failed to understand or control many things. We still do, but at least we now appreciate that most things probably remain explicable and manipulatable by natural rather than supernatural means.

Supernatural ideas do not really allow us to understand how the world works or how to control it, but they do provide two big payoffs, they give power over others to those who promulgate them, and they provide endless excuses for the inabilities of the powerful.

How come god-kings and monarchs anointed by gods, let alone priests, cannot even control lightning or heal the sick?

Thus, religions have usually taken a rather dim view of science and natural magic as they tend to encroach effectively on territory that religions had designated as supernatural.

Historically, magic has often borrowed supernatural ideas from religion because religious terminology tended to dominate many cultures' entire world views, but magicians have generally sought to understand and control rather than supplicate the various spirits set over natural phenomena.

Religions have never shied away from asking favors from supernatural sources, most priestly activity and prayer consists of little else, but they do not like to think of this as magic, and if, as often usual, the request fails, then obviously the supernatural forces know better.

291) What historical moments or trends do you think influenced your childhood the most?

WW2 ended seven years before my birth in 1953, rationing of some items continued till about then apparently, a lot of people look back on the postwar period in the UK as rather dreary times until the 'swinging sixties' came along. Everything in Britain tended to run on class-based lines, I did not perceive the times as dreary as I had nothing to compare them with. The working classes only read about the 'swinging sixties' in the Sunday papers, it didn't happen for them until later in the seventies.

WW2 cast a long shadow over my childhood. Most boy's comics featured little else, and most model kits consisted of WW2 tanks, aircraft, and ships. The British seemed to feel that WW2 represented their finest hour, but patriotism evaporated with the realization that they had won a pyrrhic victory and only with enormous American help, and that it had cost the nation its empire and left it in terrible debt. Postwar Britain did not become 'the land fit for heroes' that my parents' wartime generation had been promised.

American culture had a considerable impact on my childhood. Marvel and DC comics provided entertainments and imaginations that did not hark back to WW2. Rock and pop music formed a backdrop to my teenage years, quite a lot of that came from British sources but I guess only because that had developed from the American model of teenage culture.

I suppose postwar educational initiatives did have a large influence on me, particularly the newfound status of science in education. In my grammar school, fully half of the boys studied mainly science for the last two years, in previous generations the liberal arts and classics had held much more importance.

A lot of historians like to look upon the events of 1968 as a turning point in western culture, we had massive protests over the Vietnam war, the huge student riots in Paris, and the beginning of the Hippy movement. For the scions of the British upper working and lower middle classes the new ideas caught on from the early seventies onwards, just as I entered my formative years. Youth culture prior to that seemed limited to the teddy boys, mods and rockers, and the skinheads, all of whom celebrated teenage violence and various exaggerations of adult culture and attitudes.

The Hippy movement experimented with values quite opposed to that of its parents' generation. We advocated free love (the contraceptive pill had just become widely available), we took hallucinogens like marijuana and LSD rather than the mere mood-altering substances favored by our parents and the preceding youth cultures. We declined to enter the conventional professions or to dress for them. A lot of us lived as cheaply as possible (social security seemed generous at the time, and cheap or free housing remained plentiful. A huge interest in esoterics, eastern mysticism, and occultism developed. Identity goals became more important that economic ones, some people 'dropped out' and went to live in communes or teepees. Some of us went the whole hog and hitch-hiked overland to India in search of ourselves or enlightenment or simply adventure and different cultures.

292) What do you think made you the right person to birth chaos magic and what do you think most influenced those parts of you?

I happened to come along at the right time with the appropriate skills. If I saw far, I did so because 'I stood on the shoulders of giants' as Newton quipped.

I have come to recognize Samuel Liddel 'MacGregor' Mathers, 8 or 11/1/1854 to 5 or 20/11/1918, (we may well share a birthday), as the main giant on whose shoulders I stood.

He came from humble origins but became a highly skilled ancient linguist and he seems to have singlehandedly synthesized the entire corpus of the Golden Dawn system out of the confusing mass of classical and medieval and renaissance and eastern esoteric thinking.

He made some messy compromises with eastern traditions of hidden masters and freemasonry and forced everything to fit somewhat uncomfortably under the umbrella of a Pagan-Monotheist Neo-Platonic Kabala, but he created a reasonably coherent system that well suited the Romantic Revival era and formed the basis for the esoteric thinking underlying modern Wicca, Druidry, Hermetics, and the work of Aleister Crowley.

Following the lead of Schopenhauer and Levi who emphasized the supremacy of the intent and imagination of the magician, Mathers played fast and loose with the traditional symbolism and theology and the implied metaphysics of magic to create an eclectic way of thinking that acted as a precursor to Chaos Magic.

We shall perhaps never know how much his wife Moina contributed to Mathers' ideas and inspirations. They practiced magic together from the beginning and she continued his work after his death. As the sister of the celebrated and Nobel prize winning philosopher Henri Bergson, she may well have had a familiarity with his radical ideas about immediate experience, intuition, creativity, time, and free will.

Austin Spare had studied with the GD and Crowley, and he realized that you could strip away most of the ritualistic practices from the GD system and substitute the contents of your own subconscious for a symbol system. Crowley liked to retain the ersatz ancient symbolism and freemasonic ritual practices and augment them with sex and drugs and other tantric practices.

Then I came along with a scientific perspective in an era in which relativistic and quantum insights had partially combined with an esoteric revival in a basically irreligious culture.

The equation of Intent + Imagination + A Chaotic Universe = Magic, seemed like the next obvious step in the evolution of magical thought to me.

293) Do you expect anything specific from the next revival?

A lot of contemporary occultists still use the paradigm of the 1880s revival that Mathers created, some use the late 20th century paradigm which in its hardcore form means Chaos Magic. These probably still have to fully play out.

Based on the Psycho-historic Mechanism of the Aeons which appears in the Aeonics chapter of Liber Kaos; waves of Materialist, Transcendental, and Magical thinking interact to create the underlying philosophies and metaphysics of cultures as they evolve.

Currently in the developed nations, Materialist thinking has risen to ascendancy, perhaps to its zenith. Transcendental thinking has declined markedly, and Magical thinking in both its covert and overt forms, continues to rise.

Societal changes tend to drive the waves of paradigm change. The first magical revival had its roots in the clash of paganism and monotheism in the first and second centuries AD as the Roman empire caused massive mixing of peoples and cultures and ideas in the Hellenic world.

The Renaissance in Europe saw an apparent upsurge in magic, both ceremonial magic involving the conjuration of spirits, and natural magic which shaded into proto science. However, this does not seem to represent new thinking, but merely a rediscovery of Hellenic ideas largely lost or suppressed during the dark ages.

The second magical revival beginning in the 1880s had three main roots, a romantic revival in reaction to industrialization and mechanistic science, and the eclectic perspective derived from colonialism's interface with oriental cultures.

The third magical revival owed as much to the decline of religion as it did to anti-materialism and, almost paradoxically, the perspectives of relativistic and quantum sciences as well.

I have no idea what societal changes may drive a fourth magical revival. Perhaps a decline or collapse of industrial societies may provoke a resurgence of transcendental thinking amongst survivors left in the ruins. Perhaps technical discoveries making mind-machine interface more effective and intimate will lead to an expansion of magical ideas. I advocate enchanting long and divining short, so I do not do futurology, yet I suspect that we may stand on the brink of an era when we discover how to seriously mess with our own biology and experiment with redesigning ourselves.

294) Do you think time travel is possible?

Change creates time. We can in principle slow down change in ourselves by using the relativistic effects of gravitational fields or accelerations so that we age more slowly whilst the world changes at the usual rate. In principle, an apparently short trip very close to lightspeed could bring you back to a world in which thousands of years had passed. A few years spent at the bottom of a deep mine might add a fraction of a second to the life expectancy you would have had on the surface.

As change creates time, 'traveling' to the past would mean undoing all the changes to the universe that have occurred since the time you wanted to go back to.

However there seems a strange problem in principle with recreating the past. The future remains indeterminate rather than merely unknown. The present moment has many possible futures. We commonly regard the past as determinate in the sense that certain things definitely happened even if they remain unknown, yet the evidence says otherwise. The past exists only now in the sense of the records and memories we have of it. Any events which could have led to this moment of now (with all its records and memories), could have led to this moment of now, so the past has the same sort of indeterminacy as the future.

These effects become measurable on the quantum scale. The behavior of a particle seems to arise from all that it might have done and all that it might do, because at the moment of now it consists of an interference pattern of waves from its past and its future.

On an 'individual particle' level this interference occurs probabilistically rather than deterministically. On the macroscopic level we do not usually notice this because the random behavior of the billions of particles that make up everyday objects usually averages out to produce fairly predictable behavior in the short term.

In the reference frame of 'now' the past and the future consist psychologically of just memories and expectations, and physically of just waves of probability, not the particulate matter apparent at the moment of now.

We cannot 'go' to either because neither has a particulate matter structure to make it a 'place' to go to. However, we can occasionally sense and interact with the probability waves of what might have happened and what might happen. The probabilistic nature of the past and the future suggests that we should enchant long and divine short, and also accept the possibilities of retroactive enchantment and forward divination acting as enchantment.

295) So then when you send an entity back in time it's relaying back a probability?

When you try to retrieve information from the past or future either directly with divination or indirectly using an entity or servitor you can only detect the relationship of probabilistic events there with the present. Hopefully, you can pick up the dominant probability if one exists. This may well apply to the past more than to the future because entropy increases locally in this part of the universe. Trying to divine forward past an event like a dice throw or a roulette wheel spin will usually prove pointless because if anything starts bouncing more than about seven times its final state will remain indeterminate. In divination the magician should seek concealed but high probability events.

I accepted the result of the mass divinatory experiment to scry the state of the universe at the supposed time of the big bang as highly probable because all participants who obtained a result got the same result that nobody had expected; that the universe looked broadly the same then as it does now. It took me a further 30 years to reconcile this result with a reinterpretation of theory and observational data.

296) Is there anything about America government that you do like?

I have only visited the USA for 3 separate weeks but my sister has lived there and raised a family over the last 30 years, plus we do get a great deal of news from the USA in UK media, and I read a lot of history.

The USA seems governed by democracy and aristocracy, with big money and big business comprising the aristocracy and having a greater proportional political effect than here in Europe. America did a lot to defend and promote democracy worldwide during the 20th century but it made some rotten compromises with bad regimes around the world just because they opposed communism.

The American doctrine of economic freedom has made it the wealthiest per capita of all large nations on earth, and also one of the most creative and heavily armed.

Democracy always lies vulnerable to mob rule. Aristocracy always lies vulnerable to oligarchy. For the last 30 years the American professional and managerial classes seem to have become more oligarchic, more protective of their own status, and greedier. The American rejection of any sort of monarchical principle probably prevents tyranny, although a populist elective mob rule dictatorship remains a possibility if the system continues to marginalize the working classes.

The American state provides far fewer social services than here in Europe and this probably accounts for the continuing community function of religion there. Yet the complete secularization of public education in the USA has probably led to less social divisiveness on purely religious grounds.

The American system of work-related health insurance looks like the triumph of entrepreneurialism over humanitarian principles to us Brits. We have a state funded system but it suffers from appallingly bad management which makes it expensive and inefficient. Yet life expectancy in the USA has fallen below that in much of Europe.

In the medium term I look forward to seeing how America responds to, and compares with, its new main rival China. How will individual freedom and capitalism stack up against state-controlled capitalism? Europe seems rather in the middle with its mixed economy doctrines. I wonder if either extreme will find it necessary to move a little closer to the middle.

On balance I prefer the American system of government to just about anything outside of Europe, India, and Australasia, but I think all four of these polities can learn something from each other.

297) Do you think the US or China have more of an advantage dealing with climate change?

'Dealing with climate change' implies two things, trying to reduce it, and dealing with the effects of it.

The Chinese Communist Party (CCP) leads an authoritarian command economy in which it can direct agriculture, industry, labor, and population distribution by fiat, but only up to a point. Its mandate hardly has any ideological justification today and now derives entirely from its delivery of social order and economic growth and a bit of nationalism. If economic growth falters the CCP may lose its grip and the system may implode as it has done many times in the long history of China. Leadership cliques and millions of people usually die in these periodic paroxysms.

The CCP will resist doing anything that imperils the economic growth that keeps it in power, but it will not give a damn about the effects of its policies on minorities in its own territories or on the rest of the world.

The Chinese system however looks better placed than in the past to move to new technologies, particularly if they prove more economically efficient, and it looks better placed to deal with localized agricultural failure and famine arising from climate change.

In the US, the government has far less power of coercion by fiat, and in a democratic/aristocratic system it comes under intense pressure from the vested interests of vocal minorities, big money, and big business, yet it can change administration without revolution and executions.

Humanity's prospects of ameliorating climate change and averting catastrophe depend very much on making alternatives to fossil fuels cheaper than the fossil fuels, either by technical innovation or by punitive legislation, or a bit of both.

If economic growth reverses, either from the direct effects of climate change or from policies designed to ameliorate it, then the Chinese will probably respond by repressing their poor and the Americans will probably respond by repressing their rich, because both have rather too many of each.

298) How much of our time perception is due to the way the human nervous system processes change? I think I heard somewhere certain animals have a different time sense?

I doubt that we have any perception of time at all, we just have a conception of time based on the perception or anticipation of change. It seems fruitless to inquire what time 'is' or even what time 'does' because we have nothing to observe or theorize about except relative rates of change. We define cesium atomic clocks as our currently most accurate only because if we set two of them going and count their vibrations electronically then for days and years to come, they will both show exactly the same enormous number of vibrations when we observe them. From this we infer that the 'lengths' of days and years themselves do vary slightly, at least relative to cesium clocks.

Reaction times and perception times vary a lot between species. You cannot fool an octopus with an ordinary television picture of its prey or predators. For that you have to use high-definition television, otherwise it perceives the line scan which moves only fast enough to fool us. No human skydiver could snatch a sparrow from the skies in a dive at 120 mph, as a Peregrine falcon can. Houseflies usually evade our attempts to catch them with our bare hands.

299) To what degree do you think we can mess with our time sense?

Our sense of time remains subjective and variable; excitement, danger, pain, pleasure, and boredom can all make a big difference not only to the experience of time, but also to the memory of it. Some forms of shock seem to subjectively dilate time, a terrible moment can seem to go on forever, but other forms of shock seem to leave holes in the memory.

300) What are some ways one might attempt that?

Falling asleep usually obliterates time for me. Paying serious attention to either my own breathing in meditation, or to the natural environment when out walking seems to subjectively slow down time. On the other hand, after intense intellectual activity such as reading something difficult or writing or mathematics, I become surprised by the elapsed external clock time.

301) Can that be of any use to magic?

Quiescence seems to favor divination. Excitement seems to favor enchantment. Perhaps both allow the magician to somehow do more than normal in the clock time available.

My ongoing calculations and speculations concerning the quantum realm suggest that a lot of what we call magic involves reversible time and a plane of 'imaginary time' that acts as a kind of pseudo-space, and these underlie what some occultists have called the 'astral plane'.

302) Which of the Abrahamic religions do you think allows for or spawned more mysticism or magic?

Scholarship has opened a vast area of study here. We also need to consider Zoroastrianism, the religion of ancient Persia. The seventh century Moslem Arab conquest of Persia almost completely obliterated Zoroastrianism but it had an extraordinarily rich tradition of magic, indeed the words Magic and Magus come to us from the Magi, the priests of this very ancient religion. Zoroastrians believed in a great cosmic dualism of a good god versus a bad god or devil. The Israelites originally had a polytheistic system and their hardline monotheism may have developed or hardened in response to the captivity of their elite in Babylon around 600 BC and they probably picked up many magical ideas from the Persian Magi. The biblical Exodus of the Israelites from Egypt appears to have become concocted around the fifth century BC as a national foundation myth and has little basis in archaeology or historical scholarship. Ancient Egyptian magic seems to have influenced other cultures only after the Roman conquest of Egypt in 30 BC when many Egyptian temple priests forced out of business became itinerant magicians.

All religions develop syncretically, they do not spring full-blown from a single revelation. The Israelite's polytheism developed into the Judaic monotheism by a process of elimination perhaps aided by Zoroastrian inputs. Both Christianity and Islam took many ideas from Judaism and all of them tended to retain previous pagan deities as devils and demons.

All three Abrahamic monotheisms have at various times had very extensive mystical and magical traditions accompanying them. The guardians of religious orthodoxy have sometimes accepted these traditions with caveats, at other times they have adopted a completely hostile attitude. The monotheisms cannot deny the possibility of magic without calling into question the supernatural basis of their faiths, but they always remain keen to keep it under religious control and tend to forbid and persecute any form of magic that does not devolve from the power and authority of their god.

Mystics can create political problems for monotheisms. Independent enlightenments and revelations can threaten the power of religious authorities and the relationships between religions and secular authorities. I suspect that, historically, monotheist religions have exerted far more effort on suppressing rogue mysticism than on suppressing rogue magic. Religious wars usually have additional political and economic causes, but a religious war always implies the suppression of someone else's mysticism.

All three monotheisms exhibit the same themes in their magical traditions, any magical idea that you can find in one of them you can find in both of the others. They all had the equivalents of angels, demons, and spirits, and similar neo-platonic ideas about natural phenomena and natural magic. All three cultures seem to have exchanged esoteric ideas and manuscripts rather freely over the centuries. However, the historical Islamic material often proves more difficult to access due to cultural developments in the last 500 years.

The Persian Al-Ghazali (AD 1058 – 1111), wrote 'The Incoherence of the Philosophers' which became hugely influential in the Islamic world. He rejected the study of ideas from the classical Hellenic period, he rejected intellectual curiosity for its own sake, and he advocated a form of strictly Islamic Sufi mysticism. The growth of his following over the succeeding centuries appears to have done much to suppress interest in both science and magic. Following the end of major Islamic expansion in the sixteenth century, both its scientific and magical traditions seemed to go noticeably quiet, and mysticism became ascendant. In Judaic and Christian cultures meanwhile, magical, and scientific enquiry forged boldly ahead despite frequent religious opposition.

I get a huge amount of magical and scientific correspondence from people with a Judaic or Christian cultural background, but very little from those with an Islamic cultural background. Yet curiously, I have seen a few modern style cosmology papers complete with relativistic mathematics from Iranians in recent years.

303) So, I feel across the English-speaking world young people are embracing slang, due in part to the internet, but the youth of the UK seem to have a particularly fluid relationship with the meaning of words. Can you give me any insight into that?

The English language has probably the largest vocabulary of any language because it has so many roots, the old Brythonic languages, Latin from the Romans, Germanic inputs from the Angles, Saxons, and

Jutes, Norse from the Vikings, and finally French from the Normans. Thus, we have ended up with several words for most things. As the Normans conquered the Anglo-Saxon kingdom, French derived words became regarded as posher whilst we retained the Anglo-Saxon for coarser speech and expletives as in the 'F', 'C', and 'S' words. All subcultures tend to evolve their slangs and regional dialects. In my youth I could barely understand common speech and idiom from the further reaches of the British Isles. That has diminished markedly due to the media of radio and television. Today the media allow for the rapid spread of subcultural slangs and you can hear the same memes from Inverness to Cardiff.

Slangs develop to reinforce sub-cultural identities and they seem to have a fractal nature; my own family has idiosyncratic words which we do not use outside of it.

304) PC culture has also gotten really hardline about words one isn't allowed to use, what do you think the full effect of those taboos are? Are we giving these words unnecessary power?

I find it hard to keep up with the cascades of euphemisms. I have heard that some Homosexuals do not like that word anymore, perhaps Gay may soon fall out of fashion too, perhaps Poof will come back in. I really cannot remember the latest PC for dimwit. Retarded? Educationally sub-normal? Educationally challenged? Learning difficulties? Special needs? Neuro atypical? Every new euphemism tends to become regarded as an insult after a while. I remain uncertain about the current status of the terms 'person of color' and 'colored person'. I wonder if the words 'black' and 'white' will become eliminated for use in any context whatsoever, as some seem to demand.

Identity politics, non-binary-ism, woke-ism, cancel culture, culture wars, and the attempt to damn our entire history because some long dead people in it did things we now disapprove of, all seem symptoms of a society that has too much of everything and yet expects infinitely more despite that it has become bored and irritated by what it has. Grievance, victimhood, and virtue signaling have become adopted as social fashion statements or political platforms. I expect it will all evaporate with time, or if a real crisis supervenes.

I think I can detect a change of policy in the mainstream UK media. Multiculturalism seems to have given way to Inclusiveness. Everything now seems to have to include people of varied races, ethnicities, and sexualities, speaking and behaving in exactly the same way as ordinary indigenous Brits. We seem to have officially moved away from celebrating our differences to celebrating our uniformities.

Perhaps as a result, race relations here seem less worse than in a lot of places, despite the best efforts of the professional grievance mongers.

Interview 10

305) What's the relationship between poetry and slang?

Poetry conveys meaning and feeling, and hopefully more of it than in the same length of prose. Slang can complement poetry if most people understand its meaning, particularly if the slang says something more about the subject than the conventional terms do.

Shakespeare invented a great number of new words and memorable phrases; they would have qualified as slang at first.

I love the word 'Phenomenize!', it goes into many of my incantations.

As a massively abbreviated form of words, Algebra qualifies as a sort of slang, and some simple elegant equations seem like poetry to me.

306) What's the relationship between poetry and magical thinking?

Associative thinking underlies both. Apophenia reveals the poetic likeness of one thing to another and the occult or hidden connections between phenomena. Poetic incantations engage the subconscious and usually work better than prose ones and can create connections as well as reveal them.

307) Or the relationship between taboo and magical thinking?

Things become forbidden for pragmatic reasons. Experience shows that keeping pigs in hot dry countries or practicing incest eventually leads to undesirable outcomes. Such pragmatic taboos tend to become overlaid with additional religious and magical reasons over time.

Religions, sciences, and secular cultures all have taboo ideas that people should not think, speak, or write about, for the pragmatic reason that it might upset the consensus applecart.

When someone overcomes a taboo and creates a workable idea, we either persecute and/or hail them as a religious prophet, a scientific genius, or a social or political revolutionary.

Magical thinkers seem naturally drawn to antinomian thoughts about most things as they already find themselves in the taboo regions of religion and science by considering the possibilities of magic.

Some magicians find the act of breaking taboos empowering in itself; and they sometimes go too far. As a general magical rule of thumb, do not break a taboo until you have found out why it exists.

308) Let's say you were some kind of Dr Who monster, what could you harvest from humans other than meat?

I seem to remember The Doctor saying something to the effect that the universe contains very few genuinely evil creatures, most of them are simply hungry. I hope he got that right.

Anything that can travel freely around the universe would probably have no need for minerals, Lebensraum, or energy resources, or even slave labour. If it has got itself across interstellar space, then presumably it will have curiosity and a hunger for knowledge. I would hope that it would find us interesting enough for anthropological investigation and harvest our primitive culture, even if our technical prowess looked laughable to it. Perhaps we might get a handful of hyperdimensional baubles in exchange for the contents of the Louvre, but it might simply discretely take holo-solid duplicates from orbit.

Any Dr Who monster always seem amazingly advanced in some ways and ridiculously primitive in others, a metaphor for us perhaps.

However, if it concludes that we will soon discover how to travel around the universe as freely as it does, then we may have an existential problem because of its probable evaluation of our behaviour.

Regretfully it may decide to drop a planetary demolition charge on the irredeemably psychotic and destructive apes on Sol 3, so that their madness does not spread.

We breed uncontrollably, we exterminate thousands of other species per year, we trash our own environment, we have billions of weapons pointed at each other, and we entertain fantasies of taking war to other worlds. Dr Who monsters – that sounds like us.

309) So, a lot of new age types like to talk about vibrations and geometry holding secrets to the universe. Is there anything there from a Magic perspective? Is it pointing at anything cool or can that sort of thinking be useful with enough agnosticism?

Esoteric thought and terminology can often seem rather analogical and ill-defined and sometimes absurdly speculative, but therein lies part of its virtue – it allows for thinking outside of religious and scientific boxes and it probes the squidgy edges that these boxes have.

The scientific idea of 'vibrations' has proved attractive to occultists because of the associated idea of sympathetic vibrations. One vibrating structure can set off sympathetic vibrations in a similar structure. This seems to lend weight to ideas about sympathetic magic. Magical thinking may eventually prove justified here, if quantum waves really do underlie the structure of reality and if quantum waves consist of vibrations of spacetime, then everything consists of vibrations and all physical laws, and all magical phenomena, depend on sympathetic vibrations. Physics then depends on reliable forms of sympathetic vibration and magic tries to deal with the less reliable and more mysterious forms of sympathetic vibration.

Geometry fascinates many esoteric thinkers because it reveals connections between things, and it reveals scale-independent relationships. Vibrations must consist of geometries. Spacetime itself has a geometry as does the entire universe. 'Sacred' geometry attempts to express metaphysical principles by shape. Freemasonic and Enlightenment thought sought to portray the ultimate nature of the universe as a geometric phenomenon. I tend to think that we do not really understand anything until we understand its underlying dynamic geometry. We should not regard geometry as mere static shape; time also forms part of the geometry of phenomena. It seems that either the universe and all its contents consist of dynamic geometry or that constitutes the most fundamental understanding of it that we can achieve. Yet as good scientists and good magicians we should remain agnostic, the more we seem to know, the more we expand the horizons of what we know we don't know.

310) What are the practical considerations and calculations you make when deciding on a task for the Knights of Chaos?

During the period of its last active campaigning season some years ago, we tried to block several ecologically catastrophic ventures involving oil exploration and exploitation. A gratifying large number of technical failures, accidents, delays, and cancellations followed. Protest movements sprang up in many places, and the fossil fuel issue became more widely acknowledged. We used shared maps and Google Earth for targeting the specially designed servitors during online rituals.

We also conjured for changes of attitude about and within one of the world's largest religions and I combined this with a trip I had planned to Rome for other reasons. Within a short period, the amount of material in the media on the terrible crimes of catholic institutions seemed to multiply enormously. A fiercely reactionary Pope resigned for the first time in history and became replaced with a surprisingly intelligent and liberal one.

With magic you pick your objectives from amongst possible futures and give them your best shots, but you can never know how much difference it made.

Discussion then turned to another of the Abrahamic monotheisms, but we could achieve no consensus on what we wanted to achieve or even about whether we should attempt to do anything. At the time, the unintended consequence of foreign interventions loomed large in many minds. The Knights and Dames of the KoC have since rested their wands, and online Arcanorium College where we held our round table eventually closed due to the exhaustion of the online seminar format.

311) But how do you pick targets, mitigate unanticipated consequences, find the attack point, etc?

All the Knights and Dames built themselves an armoury of servitors to the same design and each had its own evocation and launch ritual and purpose. With servitors we aimed to give the evoked entities the freedom to try and create optimal outcomes. Having used our collective intelligence to build them, we launched them and hoped for the best.

312) Tell me more about the seminars?

The Maybelogic Academy that Robert Anton Wilson's friends set up, invited well known figures to give courses lasting for a month or two and then charged interested people a fee to participate. The profits supported Wilson's end of life care. I ran a couple of courses on Chaos Magic and another on Chaos Magic in Business. I would put up study material and work for participants and they would reply, and discussions would ensue. I had about 200 registrations for the three courses in total.

Quite a few participants wanted to continue indefinitely, so I set up Arcanorium College with a cheap subscription to cover the site maintenance and modest expenses for guest tutors. I invited a wide range of friends and acquaintances who had made something of a name for themselves in various magical and esoteric fields and invited them to give six week courses on their specialisms and interests. I gave quite a few courses myself over the years and established the KoC on the site.

It all worked brilliantly well for some years. I wrote both The Apophenion and The Octavo as a result of the inspirations that arose from it. Participants contributed very actively, and we often had more than thirty posts to read and discuss each day. The archive of the site became huge.

As the years went by, tutors ran out of material, participants began to want continual innovation and novelty, and their attention spans began to shorten, few seemed to want to concentrate on a theme or a schedule of work for long. I do wonder if the internet format itself tends to downgrade the quality of communication eventually. Today the net seems to have a lot of occult forums where participants just seem to shout their opinions at each other, and measured debate has gone out of the window.

313) Are there recordings of the seminars?

No, according to my IT consultant, the physical manifestation of Arcanorium College will have consisted of two cigarette packet sized data storage modules, one in a bombproof bunker beneath somewhere in southern England and a backup duplicate in another bombproof bunker somewhere beneath southern California. As soon as you cease paying the rent on them some minion comes along and pulls them out of their slot, wipes them clean and re-inserts them for use by another paying customer. Tens of thousands of hours of correspondence and thousands of images disappear in a second.

I do wonder if future historians will have a severe shortage of material to work on for the twenty first century. They may have almost no idea what anything or anyone looked like, or what anyone said or wrote to anyone about anything. All the peta-bytes of data on all the social media platforms and websites seem destined for deletion within the lifetimes of the people who put it there, or formats and technologies will change. Books, letters, and photographs may become rare and treasured relics as never before.

314) Can you use magic to talk to plants and animals?

Humans can certainly communicate with plants and animals, but except perhaps with higher animals, the talking part seems mainly for our benefit.

My wife has become a highly skilled horticulturalist and has created an extraordinary series of fruit, vegetable, and flower, gardens containing many hundreds of species around our home. She will often say, that does not seem happy there, it wants more shade, or this plant does not like that one, so I'll have to separate them. People say that she has 'green fingers', but she also has 'black' ones. When she takes a dislike to a plant, as she has done with some of the vegetation left by the previous owners of the house and grounds, it tends to mysteriously die. I mostly just do the donkey work in the gardens although I have spent eight years raising a trio of Mandrakes from seed that a Greek Magician correspondent sent me. I do talk to them, and I seem to have developed enough empathy with them to at least keep them alive and thriving in the chilly British climate which they dislike.

Pre-agricultural and pre-literate people often have astonishingly detailed knowledge about what they can and cannot safely eat, and what acts as medicine or poison for what. Considering just how many thousands of species they had to choose from, even in temperate climates, let alone jungles, it seems hard to believe that they worked all this out by trial and error or inference. Most psychotropic plants bear little resemblance to edible ones.

Civilisations arose when we domesticated cereal crops, but you can argue that those cereal crops effectively domesticated humans and made them spread their seeds over vast areas and defend them from predators and competitors. Osiris, like many of the classical gods of death and rebirth seems to have begun life as a deity of wheat which grows and dies and becomes reborn. The ancient Egyptians venerated Horus the falcon and Bast the cat as deities because they defend wheat and its seed from one of its main enemies – rodents.

Theriomorphic deities in pagan pantheons the world over seem to have developed from preceding Animist and Shamanic cultures which considered everything animate or alive in some sense, not just animals and plants but also mountains and rivers and other natural phenomena.

I do not generally like domestic dogs but in India I formed a relationship with a stray dog up in the mountains of Himachal Pradesh and another who hung around our boatbuilding camp in Goa. Strays in India have evolved all sorts of special skills to survive. Both would show up if I merely thought about putting out some food out for them, either that or they put the thought into my head.

315) You have a pointy hat and I imagine a few other pieces of ceremonial clothing; how do readers imbue their own accessories with ritual or magical significance?

In principle you can do all forms of magic empty handed. Yet I find that about as difficult as doing theoretical physics entirely in my head without pen, paper, diagrams, or a pocket maths calculator. In practise I like to have a lot of physical supports that I have made or designed myself. Over the course of a career, I have redesigned my wand many times, the current main version in pearwood grown in my own gardens and bearing nineteen carved symbols may represent its final form. The nineteen symbols and their arrangement relative to each other represent many complex concepts all condensed into a device that fits in a deep jacket pocket.

I make my own servitor ground-sleeves and deity-forms for invocation, in a variety of materials, sometimes exotic hardwoods, sometimes in metals that I have either cast myself or carved the wax originals for. Occasionally I have used composites and carving for larger altar sized pieces. I need to develop intensely visualised images of all these forms and making them personally helps a lot. In some ways making a servitor ground-sleeve represents an act of command and making a deity statue represents an act of devotion.

Magical clothing serves several purposes for me. Putting on a robe and pointy hat in the privacy of my conjuratorium marks a break from ordinary life and a concentration on something of intense significance. This also applies to group work when I participate in it. For everyday dress I usually wear a semi-discreet amulet, a mildly pointed modified hat and carry a concealed pocket wand. These remind me of my vocation and not to waste too much time on ephemeral matters. Sometimes they provoke interesting conversations with those having the wit to recognise them.

316) What do you think about Korzybski "time binding" and George Anderla's ideas about the doubling of information?

Language and writing have certainly allowed humans to preserve or bind information across time and across many generations in a way that no other species on this planet can, and this underlies our cultural history. Writing has such power that the ancient Egyptians ascribed its invention to their magician god Thoth.

The Jewish, Christian, and Islamic religions survived and spread around the world because they all depended more on a portable book than on immobile temples and the memories of priests. Such books had to become defined as 'sacred' and 'infallible' to perform this function. Written evidence has carried more clout that verbal evidence ever since.

The Royal Society formed in 1660 and adopted the motto 'Nullius in Verba', nothing verbal, we do not accept anecdotes, we want everything in writing. It became the world's premier scientific institution; it established the scientific method and kickstarted the Enlightenment and the Industrial Revolution.

The Korzybski school of general semantics, via Robert Anton Wilson, came to influence my thoughts about the concept of 'being' as a false concept. You may have noticed that I have avoided using the 'is' word throughout this series of interviews. Nothing has 'being', all events and all phenomena consist of what they do, including us. Even a rock consists of a dynamic process, its constituent particles continuously interact with each other and with the rest of the universe. When we say that anything or anyone 'is' something, we commit an act of sloppy shorthand thinking.

Anderla may have a point about the doubling time of recorded information shortening exponentially as history currently progresses. Yet if information only exists if something or someone can interpret it, then perhaps we should define knowledge as information that conveys something meaningful and useful. The Babylonians have left us a wealth of inscribed clay tablets, most of them just debt and tax records. Just how meaningful and useful the tsunami of recorded information we create today will appear to future generations remains an open question. I doubt that we shall even bother to preserve social media records for long. The discovered 'facts' of science may seem to multiply and may seem destined for preservation, but the half-life of such facts also seems to shorten as do the opinions which masquerade as facts in the social sciences and liberal arts. We seem to discard knowledge almost as fast as we create it. A thirty year old textbook on almost any subject now looks like an embarrassment. I doubt that the historians of the future will credit us with creating meaningful and useful lasting knowledge at anything like the rate we imagine.

317) What's the basis of, and is there anything to, the idea that information is the unifying structure that allows for magic and morphogenetic fields and particles maintaining relationships and all that?

Information always appears in embodied form, carried by some medium or other, it doesn't seem capable of crossing space or time on its own. Quantum entanglement may seem to violate this principle but not if you use TIQM, the transactional interpretation of quantum physics. In this interpretation, time reversed waves travel back from the receiving event to the emitting event of the entanglement, creating the appearance of non-locality and instantaneous effect.

On the other hand, the underlying fields through which the waves travel must have a non-local and instantaneous effect to keep gravitational and electrostatic fields precisely aligned with fast moving objects and to maintain the observed uniformity of physical laws and constants across the cosmos. (Note that this currently remains an unfalsified heresy.) It would seem not unreasonable to suggest that such fields may carry some information about the structure of the matter which subtends them.

318) Do you think there are macro physical conditions in the universe that are more conducive to magic?

The above two examples, entanglement and non-local fields, imply that the universe has far more connectivity than simple material causality suggests.

Additionally, if as I suspect, the universe has a finite and unbounded (hyperspherical) spacetime geometry then intelligent life has had, and will have, effectively unlimited time to develop. Thus, the universe will contain intelligences vastly more knowledgeable than us. Some of them may reply if we ask the right kind of questions.

319) Like if you're near a black hole and time relative to the rest of the universe is slowed or sped up or whatever, could that affect your probabilities and your outcomes?

Anything under a strong acceleration or equivalently in a strong gravitational field passes through time more slowly than anything under a lesser acceleration or in a lesser gravitational field. The effect however remains miniscule at the sort of accelerations and gravitational fields the human body can survive.

Perhaps the starship crews of the future, if we get that far, may discover that when their vessels warp spacetime they also warp probabilities and outcomes in peculiar ways. I did once start work on a psi-fi novel with this highly speculative what-if theme, but I lost the plot on it as it became too complicated. This thought experiment at least led me to reject all versions of Multiple Universe theory. Waves can have a multiple and probabilistic nature, but particulate reality must remain singular if the universe has any conserved properties at all, and it does seem to have quite a few.

320) What do you think happens when we die?

Plainly the physical body ceases to function and begins to disintegrate, to eventually become recycled at the atomic and molecular level into other bodies. Statistically we should all contain a few particles that once formed part of the body of Julius Caesar, or of any other historical person. The sense of personal self seems little more than an illusion created by our nervous systems for good evolutionary reasons – it adds survival value to the genes which give rise to it, yet even the most superficial introspection reveals that the sense of self does not remain constant and unchanging from cradle to grave and neither does it operate during deep sleep. If our sense of self or 'soul' consists of anything it must consist of memory or the formless experience of Kia, - aliveness, yet these two also seem to fall apart at death, so if we posit an afterlife or reincarnation, we must wonder precisely what could survive into it.

Some of our ancestors seem to have practised funerary cannibalism, figuring that the dead would prefer to physically reside within the living tribe rather than in the cold earth or in the guts of bugs and scavengers.

Similarly, we have some options for the fate of our information content. It can form part of a collectively imagined afterlife or part of the collective imagination of incarnate life, but in either case, preserving coherent memories and any sense of personal self requires more effort than most people feel like committing in this life. Personally, I remain content to leave descendants who carry some of my physical patterns and thoughts, and several books which will allow me to still speak inside people's heads. I shall willingly surrender to a chaotic reshuffling and redistribution of the rest of me in the sure knowledge that it will enrich surviving humanity.

321) Do you think near death experiences can give us glimpses of anything meaningful about reality ?

Reported near death experiences seem highly variable and subject to cultural expectations. They probably tell us more about those cultural expectations and neurophysiology than anything else. Rare instances of 'Remote Viewing' by the almost dead, and remote viewing of the just dead by the living, do however raise interesting questions about parapsychology. I have had a couple of personal experiences of the latter myself.

322) What do you think are the most common misconceptions about chaos magic or magic in general?

Firstly, concerning magic in general, there seems a common misconception that humanity proceeded from magic to religion to science. A closer analysis reveals that magic gave birth to both religion and science and that all three ways of thinking remain highly active. The original set of magical ideas appears as Animism, the theory that all natural phenomena, plants, animals, and people have some sort of 'mana' or 'animating force'. This idea survives explicitly and formally in the theory of Panpsychism and somewhat informally in the idea that phenomena have essences. In the classical Hellenic world, it took the forms of Platonism and Neo-Platonism in which all phenomena have underlying and immaterial 'Forms'.

Shamanism, which makes extensive use of trance states to interact with the 'animating forces', developed as the main expression of animism in northern latitudes.

Animist ideas readily lend themselves to elaboration into ideas about 'spirits' and 'deities' that people need to placate or bargain with, and all religions grow out of this. Such religious ideas also feedback into magic to create the sort of learned magic in which magicians invoke and evoke and command various spirits to achieve effects.

Animist ideas also lend themselves to the elaboration of ideas about precisely how the animating forces work, and all science follows from this.

Magic in the popular imagination often seems to mean doing the impossible, breaking the laws of physics, and an imaginary world that has no fixed rules at all. I always find fictional magic of this style rather irritating and of poor entertainment value. The feeling that the authors of such fictions will allow anything to happen makes me cease to care what will happen. In practise you cannot break the laws of the universe, but you can bend them to achieve highly improbable but not impossible outcomes. The laws of the universe arise from chaos and work by statistical probability not certainty. Science attempts to understand and manipulate events of quite high probability, repeatability, and reliability. Magic concentrates on events at the other end of the spectrum.

Chaos magic seeks to exploit the randomness and indeterminacy within statistical probability. The idea of making magic out of chaos came first, and then after the blossoming of Chaos Theory as a science in the popular imagination, the new tradition formally adopted the title of Chaos Magic.

This captured the imagination of many, and it infuriated many more. Chaos magic presented the 'sacred' techniques of many religious, mystical, and occult traditions as mere technical procedures adaptable to most forms of results based magic. It showed that we could work magic using any kind of symbolic system and that belief itself could function as a tool rather than as an end in itself. In doing this it acquired a sort of amoral off-white charisma, and it showed very little deference to 'ancient wisdom'. It seems rather like science in these respects, and as an irreligious science of magic it scores top marks for blasphemy in many quarters, much to my delight.

323) One of Wilhelm Reich's nuttier sounding ideas was cloud busting but I just read an article about using drones to electrify clouds to make them denser causing rain. Is this similar to Reichs model?

The American authorities eventually turned on Reich, destroyed his archives and imprisoned him. His idea about sexual liberation appeared shocking to most people in his time but have since become widely acknowledged. His theories about orgone energy and devices to enhance and project it remain matters of debate, yet he certainly got paid afterwards by the farmers who employed him to use his cloud-buster devices to bring rain to their parched fields. We can make clouds drop rain by seeding them with tiny particles that can act as nucleation centres for droplet formation. Perhaps electrical charges distributed in appropriate ways could have a similar effect. Silver Iodide particles work well but they cost a lot. Smoke has a noticeable effect; rain often falls after a widespread fire.

When out walking I cast a local rain stopping sigil quite frequently in these wet Atlantic facing islands. I haven't conducted a statistical analysis of its effectiveness, but it always seems worth a try. My biggest challenge came with my daughter's wedding in the Scottish Highlands, early in the month of May some years ago. They charged me with preventing rain. I had to wear a kilt for the occasion, so I wore a pocket wand disguised as a sgian-dubh or sock dirk, and every ten minutes or so I excused myself from the mostly outdoor festivities and went into a car park from where I had a ten mile upwind view of the clouds scudding in over the mountains. The struggle went on for hours, finally a light dusting of snow fell on the lawns as we posed for photographs and this added a delightful sparkle, but it didn't rain on us at all, despite the weather forecast.

324) What do you think the potential is for engineering weather?

We have already engineered the weather to make it more extreme and unpredictable by burning hundreds of gigatons of fossil fuels and by deforestation. We could do a lot by reforestation and by switching to renewable energy sources, but we will probably fail to do enough before the situation becomes much more serious or catastrophic. I have seen schemes for last ditch emergency geoengineering that involve spraying water vapour or sulphur dioxide into the atmosphere from tens of thousands of aircraft. This would reflect more sunlight back into space, but the water vapour may not remain airborne for long and the sulphur dioxide would eventually fall as horribly acidic rain. Even crazier schemes like the controlled opening of volcanoes and the selective nuking of the eyes of hurricanes sometimes resurface.

325) Or is it another irresponsible use of science?

Few things stimulate science like war. In the years of WW2, research and development leapt forward by about thirty years in just six. We always seem to use the products of science in a gung-ho and irresponsible fashion either because of war or because of commercial pressures to exploit inventions. In an ideal world we would not implement any innovation in a hurry, we would have a moratorium on it for long enough to assess its lasting effects. If we had spent fifty years working on the safe use of nuclear power, we might now have a worldwide network of thorium reactors that produce manageable waste, no meltdowns, and no bomb material, rather than the filthy and dangerous uranium/ plutonium systems we rushed into widespread use soon after their invention.

326) Let's say the internet went away tomorrow, do you think we would return to the way things were or do you think there would be some lasting social impact? Or would we be crippled without it?

I'll offer speculations on two scenarios here, one in which electronics in general cease to function, and the other in which 'the internet' as people use it personally for communication, entertainment, and information, ceases to function.

The first case could occur quite suddenly without a prior massive nuclear or cyberwar exchange, or a general collapse of civilisation, if the sun spits a big ball of plasma at us. That happened in 1859 and it massively damaged the primitive telegraph networks of the time. The sun spat out another vast blob of ionised plasma in 2012 but not in the direction of earth. If it had impacted the earth's magnetosphere and atmosphere it could have fried our electricity grids and destroyed most of our electronic devices. We have become so dependent on these for all forms of monitoring, control, organisation, finance, and food distribution that it seems difficult to imagine how we could avoid social breakdown, a general collapse of civilisation, and megadeaths. I doubt that we could put lower tech systems back in place or repair the damage in time to avert catastrophe and a new dark age.

If for some reason, society merely decided to turn off 'the internet' for personal use we would certainly notice many social and psychological benefits. People would have to start communicating face to face or by telephone or letter, and this would cause a vast improvement in communication behaviour and civility. A lot of the political extremism, violent pornography, abusiveness, disinformation, and fraud that thrives on the anonymity of the internet would disappear. People would learn to redefine their identities in terms of their actual social experiences rather than in terms of their largely unreal internet experiences.

The disappearance of the internet would probably initiate a renaissance in the esoteric and magical domains. Academics have informed me that surveys show that most contemporary occultists communicate with each other only by electronic means. They don't get together to do rituals and have proper meetings of minds and personalities and have deep exchanges of ideas and inspirations.

In the early days of widespread internet use a lot of high quality exchanges of material, ideas and contacts seemed to occur. Yet over time, the volume of material increased enormously, and the attention spans of users decreased markedly. Today I would advise anyone to start magical work with a few source books and a few friends and collaborators and to use the internet only sparingly for reference.

Interview 11

327) Can you give me any insight into the sexism of most of the world's religions? Taboo out of control? Corrupt institutions?

Religions grow out of cultures, incorporate a selection of their attitudes, and feed them back into those cultures with 'divine' reinforcement.

Women probably had higher social status in pre-agricultural societies. Despite the divisions of labour arising from the burdens of pregnancy and childcare and their usually slightly lesser physical strength, women's economic contributions to hunter gatherer societies probably gave them a status and personal freedom roughly equal to men.

Women probably invented agriculture through doing more gathering than hunting, and in doing so they inadvertently led the way to their own enslavement, the serfdom of most men as well, hierarchical societies, organised warfare, empires, and civilisation.

Agriculture leads to the ownership of land and people. The lord will want stable monogamous couples tied to plots of his land and he will also want certainty about the paternity of the sons to whom he will hand over ownership of that land. He will send out his thugs and priests to enforce his wishes.

The early Israelites appear to have had a pagan religion in which the main male god Yahweh had a wife called Asherah. Eventually the priests of Yahweh got the upper hand and suppressed the worship of Asherah, and their religion became monotheistic, and male dominated. These changes probably accompanied a change from a semi-itinerant pastoral and gardening economy to a fully settled agricultural one, with the women kept even more firmly in their designated places than the men.

All 'civilised' societies worked like this until the industrial revolution.

Industrialisation, and later the total war that industrialism makes possible, opened up an increasingly wide raft of opportunities for women to gain economic, military, and hence social and political power.

Of course, the process occurred unevenly and with many reverses and it remains incomplete. Some religions revise their doctrines faster than others, but most societies now move towards a fuller exploitation of the production and consumption potential of women. They have finally started to get their status back after inventing agriculture, meanwhile men begin to lose some of theirs after inventing industrialism.

Technologies drive societies. Mere politics, philosophies, religions, and moralities follow suit, usually grudgingly.

328) Can you explain this Google time crystal the internet is so excited about?

A spatial crystal simply has a structure which repeats regularly across a length of space. For example, a crystal of ordinary salt has alternating sodium and chlorine ions precisely spaced in a cubic lattice typically a millimetre or so across, although we can grow them bigger to dice sized cubes or beyond if we do it slowly and carefully.

A time crystal has a structure that repeats regularly across a period of time. Imagine a salt crystal in which all the sodium ions spontaneously exchanged places with all the chlorine ions precisely once every second, that would create a time crystal. Alternatively imagine a salt crystal that flipped regularly between a cubic and an octahedral shape of its own accord, that would also constitute a time crystal. At present though, they remain delicate structures, more complicated than salt, that survive only in carefully controlled conditions.

Time crystals do not require an external energy source to make them flip between states and so long as nothing extracts energy from them, they act like perpetual motion machines.

They may possibly find application in precision time measurement devices or perhaps in stabilising the notoriously fragile qubits in quantum computers or perhaps as data storage devices, or perhaps they will remain mere curiosities. Perhaps other weird properties or currently unseen revolutionary technologies will emerge from them. Maybe we could achieve quantum entanglement between time crystals. At the time of writing, they seem a mixture of curiosity and hype.

329) So, compared to other societal collapses in history do you think the super-rich living in high tech bunkers could help preserve science and culture or lead to some weird post-apocalyptic genetic division?

I'll attempt some futurology based on two sources, the fall of the western Roman Empire and the Club of Rome's report 'The Limits to Growth'. According to the report, first published in 1971, global industrial and food production should peak around 2030 and population will peak around 2050. After that, everything goes downhill. This all depends on the continuation of current trends and no game changing innovations. Climate change did not feature in the original report as seriously as we regard it now, but we can regard it as part of the anticipated general exhaustion of resources, it will render vast tracts of land uninhabitable and un-farmable.

Wealth only offers two forms of protection from social collapse, it enables mobility to escape to elsewhere if a safer elsewhere exists, and it can make it easier to seize power locally as larger political structures disintegrate. Otherwise, riches tend to make you a prime target.

It seems unlikely that we will have developed genetic technologies far enough to create enhanced super-people or a dis-enhanced servile class during the next forty years. Surviving organised knowledge and weapons technology will probably dominate any post-collapse scenario as usual.

The centralised western church with its network of missionaries and monasteries preserved literacy and lines of communication, and ironworking technology survived everywhere because of its decentralisation. Thus, we merely had a dark ages period and not a new stone age. If a new stone age had occurred, Europe would have become invaded by cultures that still had organised knowledge and iron weapons.

Any 'high tech bunker' that doesn't have the facilities to make its own food, power, and ammunition indefinitely has no future. Perhaps some island nations will attempt this.

Hope remains that we may yet innovate our way out of impending catastrophic collapse, but just in case, I have told my children how to make gunpowder from manure, lime, wood ash, and charcoal. Such knowledge could at least make them indispensable to a local warlord.

330) What about the super-rich escaping to space?

At the time of writing this seems a very long way off. At present we must carry almost everything we need for survival from Earth to orbit or to sterile places like the Moon or Mars or asteroids that we can reach. These places can only offer the resource of solar power currently. To carry all the survival necessities, we only have reaction thrust rockets, and despite all the fire and fury of their launches, such vehicles remain pathetically underpowered and fantastically expensive even for travel to orbit or to the Moon. Keeping people alive in orbital or lunar bases with current technologies requires frequent and enormously expensive payload dispatches from Earth. Thus, we mostly send machines into space, they don't need much life support, they don't mind going on very dangerous craft, and they don't come back.

I don't think that life off this planet will ever prove feasible unless we come up with some new fundamental physics that allows us to manipulate gravity, inertia, and spacetime.

The laws of physics may not even permit this, but some of us keep looking. I haven't yet seen anything in my equations or in anyone else's that offers much of a clue as to how we could feasibly do this.

331) China is working on a thorium reactor the news tells me, how does that change the calculus of nuclear power?

China has relied heavily on coal to fuel its capitalist-communist economic rise. This has led to a pollution crisis with many Chinese cities experiencing appalling smog and growing international pressure for China to reduce its carbon emissions. It has already started to act. As a heavily state-directed economy it can explore options that require massive investments for longer term payoffs that private capital or short lived democratic administrations may not find attractive. The Chinese seem well ahead of the curve on wind power development, and they will probably make a success of thorium power and eventually gain a substantial economic advantage from doing so.

In the medium term, the current Chinese authoritarian capitalist-communist system looks set to out compete democratic capitalist and mixed economy systems, but it may lack their longer term political resilience, particularly if its population starts demanding more equality and less authoritarianism.

332) What do you think the impact would be if China or Russia were to develop free or nearly free energy?

From my youth I can remember a politician in the UK saying that nuclear power would probably make electricity free by the year 2000. Recently the government here signed a deal for a huge new nuclear power station on the basis that it would guarantee a much higher price for the power it generated than what it pays for electricity from oil or gas power stations. The thinking seems to have settled on the idea of trying to go over to renewables and having a nuclear backup for when wind or solar don't deliver.

Russia already has almost free energy in the form of the huge gas reserves it sells to Europe to prop up its own otherwise crumbling and corrupt economy and gangster oligarchy.

If China develops a huge thorium power production capacity, it could export cheap electricity to neighbouring countries and thus gain more influence and leverage over them. It already props up the authoritarian regimes of North Korea and Myanmar and would probably like additional client states of a similar persuasion. A brief glance at a world map suggests some alarming geopolitical possibilities.

333) It seems to me like abstract currency that represents wealth leads to a lot of problems, do you think we'd be smart to base our currency in a practical resource?

The idea of using renewable energy capacity to back a currency perhaps deserves expanding upon. At present central banks issue debt in the form of coins, notes, and permissions which basically say, 'I promise to pay the bearer X amount'. In the old days you could take that note back to the central bank and cash it in for the appropriate amount of gold. Today they will merely exchange it for another promissory note of the same amount and probably fall about laughing.

Imagine instead a promissory notes for Kilowatt-hours of renewably produced energy. Such notes would hold their value, governments would have to invest in renewable energy capacity to achieve economic growth, that growth would be more sustainable, all other commodities would tend to become priced in terms of their real energy costs, exchange rates between differing issuing authorities should disappear, but countries which cheated would find their currencies internationally devalued. A lot of poor countries would attract energy generating investment.

334) Do you think we need currency?

We commonly consider that money exists to facilitate the exchange of goods and services. However, governments issue currencies in the first place to facilitate taxation and control. Having issued currencies, governments immediately start demanding taxes in cash rather than in tithes and services, and they always make large scale barter illegal. Plus, governments can start using cash rather than sword points to make people do things. People do not need currency as much as governments need currency.

335) Pretty much all currencies worth is based on the credit rating of their country more or less, right?

The value of any currency depends on our belief in it. As Harari pointed out in Sapiens, whilst people can exhibit belief in all manner of ideologies and religions, almost everyone believes in money today and they believe in it only because everyone else also believes in it.

Confidence in any currency (or commodity) depends on people's beliefs about how well it will maintain its value relative to other commodities. People's economic behaviour rarely conforms to the rational expectations of economists, so economics remains 'the dismal science' with almost no predictive power, but governments which do not give credible returns on the debts they issue tend to find their money going out of fashion.

The price of gold sometimes gets called 'the index of paranoia' because people invest in it when all other options seem worse. Gold has very limited utility, if we had megatons of it near the surface of the planet, we would mainly use it for roofing buildings and for dead weights, it would have few other engineering applications and even fewer chemical applications. Its rarity and chemical inertness led to its use as currency but ultimately, we had to abandon gold backed currencies because of its rarity. Historically, humanity has mined about 150,000 tons of it, nearly all of which we still have. That sounds like a lot, but now with hundreds of trillions of units of currency in play, the price of a medium sized house would correspond to a piece of gold too small to see with the naked eye if we had retained gold-backed currencies. So today all currencies have become fiduciary and backed only by the confidence that they will retain their ability to make things move. In occult terms the essence of money has undergone an alchemical transformation from Earth to Air.

Cryptocurrencies such as Bitcoin will probably fail to break governmental monopolies on currency. It surprises me that governments have not moved more firmly against them yet, perhaps they figure that it makes it easier for them to track criminal activity. I do wonder what governments do with the bitcoin stashes they sometimes confiscate. Confidence in Bitcoin supposedly derives from confidence in proof of work, but the work of Bitcoin mining consists of wasting, globally, about a small nations worth of electricity performing tortuous and pointless calculations merely to give Bitcoin a rarity value. In practise it functions as a Ponzi scheme, people invest in it hoping to make a profit from other people subsequently investing, in what has become a zero-sum game.

336) What was Crowley's philosophy on drugs? I think he was real into opium or heroin? Do you know if he tried psychedelics? It seems like he was kind of into the flirting with destruction aspect more than the enlightenment aspect?

From the vast amounts of what he wrote about himself and what others wrote about him, it becomes evident that Aleister Crowley had a gargantuan appetite for alternative states of mind. He craved excitements and would do almost anything to get out of his mind. He endlessly fell passionately in love for short periods, he had sexual relations with countless people; women, men, himself, prostitutes, and rent boys. He went in for extreme climbing and he remains the only person to have ever free climbed the towering and treacherous chalk cliffs at Beachy Head. He took just about every mind altering drug he could lay his hands on, alcohol, strong tobacco, hashish, mescaline, opium, heroin, and cocaine. He rather overdid the heroin and cocaine and had to seek treatments for addiction and nasal damage.

His mysticism and magic all revolved around his quest for attaining strange states of mind. Magic, he once quipped to Dion Fortune, 'is something we do to ourselves'. He very rarely attempted what I would call sorcery or results magic to get material things to go his way. He didn't need to, born rich he had everything he needed when younger, and after he became older and had squandered his fortune on his innumerable indulgencies, he used his considerable charisma to get people to do what he wanted.

Crowley freely mixed his sex and drug activities with his mysticism and magic, for him there was no part of himself that was not of the gods. He called what he did 'Magick', but I regard him as one of the most profoundly 'Religious' people of the 20th century, but with an unusual twist. He developed a religion in which the self becomes the only real object of devotion.

In many places he becomes quite explicit about this, in other places he alludes to it tongue in cheek, but he always remains the messiah of himself, and he saw himself as an heroic pagan god in mortal form taking his pleasures upon earth.

In defining 'self as god' he seems ahead of his time and his message found a wide audience in the post-war, post-monotheist, bohemian generations. Do what thou wilt.

Unfortunately, many who took an interest in Crowley interpreted the message as 'do as Crowley would have wilt' and made a rather poor job of trying to imitate his sociopathic behaviour. For most, 'self as god' did not prove sufficient, they failed to find their own 'True Will' and they chose to worship Crowley instead.

Austin Osman Spare had very much the same 'self as god' insight as Crowley, and I would rate him as the better magician because he also excelled in sorcery and results magic, although his writings were fewer and less easy to understand.

337) What do you think causes addiction?

A survey of chronic gamblers revealed that they had all experienced the elation of beginner's luck. Internet gambling algorithms now deliberately give beginners precisely calculated initial wins and then an improbably high number of near misses. Drugs will often initially create pleasant feelings and/or relief from bad ones. Unfortunately, persisting with either activity will lead to bad results, chronic losses of money or the redoubled return of the bad feelings after the drugs wear off and some people will keep going back to the behaviour in an ultimately hopeless attempt to regain the earlier good feelings. Interestingly, most people who receive prolonged doses of opiates after surgery do not become addicts, they may feel lousy for a few days after the withdrawal of medication, but objectively the symptoms are no worse than a dose of flu. Psychological dependency in drug addiction plays a bigger part than the direct physiological effects of the chemicals involved. The only real cure for any addiction lies in finding something else that makes you feel good instead.

338) Have you ever used a sensory deprivation tank?

No, but in the early stages of my magical career I spent a lot of time seeking withdrawal from sensory perceptions and cessation of thought with various tantric practices such as pranayama and Raja yoga. I found that these practices had two effects, firstly they brought calmness and serenity, and secondly afterwards, it brought a bursts of creativity, inspiration and occasionally psychism, as mental activity suddenly resumed.

339) Do you think remote viewing is real?

I think it sometimes works but rather unreliably as it can become mixed up with imagination and sometimes with prescience – perceiving what you will later experience. If the diviners never find out if they got it right or wrong the results seem to get worse. Plus, many remote viewing experiments ignore the magical link – if the diviner has never personally visited the target physically it seems far more difficult to visit it psychically.

340) What do you think about tales of governments experimenting with the paranormal?

Well, they remain largely tales. Both the Americans and the Russians have allegedly experimented with psychic intelligence gathering. If any government had achieved useful results, they would probably have kept them very quiet. On the other hand, if they had only negative results, they might well have dropped hints otherwise to mislead rivals into wasting their efforts. Both countries seem to maintain huge budgets for satellites so presumably psychic surveillance has not yielded much.

341) What does science have to learn from chaos magic?

Fundamentally, Religion attempts to answer the question 'Why'. Art attempts to answer the question 'Which?'. Magic attempts to answer the question 'What?'. Science attempts to answer the question 'How?'.

In practise they all take something from each other. Most scientists have some metaphysical beliefs about the 'Why' of things, they may believe in causality or 'the unreasonable effectiveness of mathematics', or simply that everything remains potentially understandable in physical terms. Most scientists appreciate the aesthetic appeal of a simple and elegant description and apply Occam's Razor in search of 'Which' idea looks the most simple, beautiful, and elegant. Most scientists like to form concepts about the nature of phenomena, 'What' powers do they have, 'What' do they do, and 'What' do they resemble.

In practise, many scientific ideas become fiercely defended as articles of faith, particularly when causes for doubt and inconsistencies between theory and observation remain. This leads to confirmation biases where confirmatory evidence takes precedence over conflicting evidence, and a path dependent theories can result in edifices built upon questionable assumptions. Science does have self-correcting mechanisms, but these tend to act rather slowly and often against passionate resistance, as with the theology in Religion, the evolution of tastes in Art, or paradigm changes in Magic.

General Relativity (with its associated cosmology) and Quantum Field Theory currently represent our most sophisticated scientific theories about how the universe works. These two pillars of scientific understanding do not include each other, rather they seem to contradict each other at present.

This perhaps arises because of an excessively rigid fixation upon the question of 'How?' in these sciences and a neglect of the secondary scientific considerations of 'Why?', 'Which?', and 'What?'.

In official quantum physics, nobody knows 'What' they are talking about. The orthodox Copenhagen Interpretation says that we do not know what really goes on in the quantum realm and that we cannot know, so just 'shut up and calculate', we can only describe it in weird maths, we can know nothing of its nature beyond that. The maths can only tell us 'How' to calculate what will probably happen. Many other minority 'Interpretations' of quantum physics do attempt to try and describe the underlying nature of quanta, but we cannot decide amongst them except upon a 'Which' or 'Why' basis, as some look more elegant or intuitive or believable than others. Thus, in the orthodox version we should ignore them all and just accept the inexplicable mathematical formalisms.

In my Hypersphere Cosmology hypothesis, I have demonstrated that if the observations which led to the current LCDM Big Bang Cosmology had occurred in a different order we could have ended up with an entirely different cosmology that fits the observations just as well, but which interprets them differently. General Relativity has several exact solutions. Historical accident seems to have determined the ones chosen for the current official model.

Path dependent epistemologies can lead to radically divergent ontologies.

Neither LCDM Big Bang Cosmology nor Quantum Field Theory seems entirely satisfactory or complete. Advocates of each theory opine that the other theory requires modification before we can develop an overarching 'Theory of Everything'.

Cosmological expansion, dark matter, dark energy, singularities and inflation all arise from mathematical extrapolations, as do all the contra-intuitive 'concepts' of quantum physics.

A Chaos Magic perspective suggests that both the cutting edges of science could profit from a little more consideration of the questions of 'Why?', 'Which?', and 'What?', because the relentless pursuit of 'How?' may have led them up mathematical blind alleys which prevent unification.

342) So, what's your stance on the Copenhagen interpretation?

The Copenhagen Interpretation remains the one currently taught in most universities. It basically eschews all attempts to interpret the maths and says we should rest content with the formulae we have cooked up to describe the quanta and predict what they will probably do.

On matters such as the mysterious double slit experiment where a single particle can act as though it passed through two slits simultaneously, and the phenomenon of entanglement where particles once in contact seem to act in a peculiarly correlated fashion no matter how far they become separated; the Copenhagen Interpretation takes an instrumentalist view. The maths can describe and predict (statistically at least) what happens; but asking why is considered fruitless.

This stance seems not unreasonable in some ways, as many interpretations remain possible, and we currently lack any means of deciding between them. However, I reject the more hardcore versions of Copenhagenism which deny the existence of any underlying more comprehensible reality.

The hope remains that some new experiment or some new input from another field will eventually allow us to sort out which interpretation applies to reality. I suspect that the Transactional Interpretation, or something like it, has a lot going for it - on cosmological grounds, specifically on hyperspherical cosmological grounds.

343) Is the new James Webb telescope going to produce any data you're interested in?

Yes, I have high hopes for it. It will look for evidence of galactic evolution within the big-bang timescale and I predict that it will not find what big-bang theorists expect. According to the conventional theory, very far distant galaxies should all appear as young galaxies that formed soon after the predicted big bang. If the experiment can overcome the difficulties of observing a representative selection of galaxies at extreme distances and putting an age to them, I expect that it will discover galaxies at all stages of development at all observable distances.

344) Are their many other cosmological models that don't believe in the big bang?

Plenty of them exist but the cosmological establishment represented by the universities and the prestigious journal publishers resists them ferociously. It took the establishment a long while to cobble together the current 'concordance' model which interprets most of the available data within a certain agreed framework and the establishment has become aggressively defensive of its model. This defensiveness arises partly because of the somewhat unscientific nature of the concordance LCDM model. L stands for dark energy, CDM stands for cold dark matter. These substances have quite fantastical properties attributed to them to make the sums add up, and they supposedly account for 95% of the entire universe, yet their fantastical properties define their existence as otherwise unconfirmable and unfalsifiable, and that looks like bad science.

345) Is that the most radical aspect of your model?

My most radical proposal hypothesises that the universe does not expand at all and has never done so. It uses one of the exact solutions of General Relativity that theorists largely chose to ignore. It explains the cosmological redshifts in terms of gravitational spacetime dilation within a hyperspherical universe that does not collapse because it has a small and potentially detectable rotation. It explains the apparent accelerating expansion of the universe not in terms of a mysterious dark energy but as an illusion created by the gravitational lensing within the hypersphere which makes distant objects appear exceptionally faint. It explains the peculiar rotational behaviour of galaxies not in terms of the presence of a mysterious dark matter, but in terms of an extra rotation caused by the rotation of the entire universe. Various parts of the model also use these new principles to get rid of the spacetime singularities and the cosmic inflation that result from the LCDM big-bang model.

346) Do you anticipate having to revise the model eventually? I feel like you said or implied things change very quickly in this field of study.

The hypersphere cosmology model does not currently specify which of several possible mechanisms gives rise to the CMBR, the cosmic microwave background radiation, although it does rule out the orthodox hypothesis that it consists of relic radiation left over from a big bang.

I hope that some observational data may emerge to allow a clarification of this issue.

It remains entirely possible that new observations may confirm or falsify either Hypersphere Cosmology or LCDM, or perhaps some other theory entirely. In terms of LCDM, the universe does seem to contain distant structures that seem alarmingly large or old, hopefully the Webb space telescope may clarify that issue.

I do find the science and the controversy in this developing field fascinating in itself, and because of its metaphysical implications. I think that what we believe about the universe at large does have meaning for our personal view of existence.

347) Can you expand on that last sentence a bit? I get the feeling lots of mainstream science dislikes the idea that cosmology has metaphysical implications.

Mainstream science dislikes the idea of using metaphysics to derive physical principles and theories, even though the principle of causality seems ultimately a metaphysical principle, as does belief in the mathematical structure of reality. When Einstein opined that 'God does not play dice' in criticism of the purely statistical predictions of quantum mechanics he spoke from metaphysical intuition. Argument still rages on this question. Some insist that the quantum realm does run on randomness, others insist that we simply haven't discovered the underlying causes (Einstein always regarded quantum physics as incomplete), some object to the use of any reference to undefinable metaphysical entities like gods, other quip that 'She can do whatever She likes.' Einstein had a Jewish background, but he never seemed to regard 'God' as a person or as anything other than a metaphor for the universe in general.

Einstein initially thought that the universe remained static and had an hyperspherical geometry and topology. A Belgian Catholic priest and distinguished physicist called Georges Lemaitre developed the idea of a universe that had expanded for what he called a 'primeval atom' and for some years an argument raged between the Einstein and Lemaitre camps about a static or expanding universe. As evidence which physicists could most easily interpret as favouring an expanding universe piled up, Lemaitre began to win the argument during the 1930s and the big bang theory became dominant. Fred Hoyle who never accepted the Lemaitre hypothesis, coined the term 'big bang' contemptuously, but it stuck.

The Catholic Pope at the time wanted to make much of the big bang theory because it seemed to offer scientific validation to the biblical idea of a creation event full of light and thunder. Lemaitre persuaded him not to make too much of it, as it remained potentially open to a falsification that would lead to a massive theological embarrassment.

Nevertheless, many monotheists who interpret scripture liberally like to regard the big bang as roughly supporting the idea of a creation event initiated by a god, and a universe with a beginning, a middle, and an end. This causes some annoyance amongst atheist scientists.

Stephen Hawking's motivation to develop a singularity free origin for the universe and a model for a universe whose boundary condition was that it had no boundary, derived partly from his atheism.

In practise our metaphysics and our physics always lie deeply entwined. For thousands of years, we thought of ourselves as living on a world surrounded by just the sun, the moon, the planets, and the fixed stars. We had no credible physical explanations for this situation, even though we developed the physics and maths to build huge pyramids and irrigation systems. We nevertheless made up many ingenious explanations, usually based on godlike versions of ourselves building the physical universe.

Now we have mechanistic explanations for the existence of this planet and celestial objects, and through evolution, our own existence. These explanations have profoundly altered our views of how reality works and of ourselves. So surely our physics has profound implications for our metaphysics, but many physicists remain wary of trying to use metaphysical ideas to search for physics ideas, although some of the best and most imaginative of them have taken some inspiration from metaphysical, esoteric, and mystical ideas from east and west.

I find the standard LCDM big bang model metaphysically

disagreeable as well as scientifically dubious because of its apocalyptic neo-monotheist linear timeframe. It all begins with an event which we cannot even in principle understand or penetrate beyond, and it will all end in either an equally apocalyptic collapse or an infinitely long entropic fade-out. Additionally, it implies that this all takes place in a literally infinite expanse of spacetime.

I cannot believe in an infinite amount of anything, we never observe it, nor could we. If the universe contained an infinite amount of anything it would have to contain an infinite amount of everything. Whenever we arrive at infinity as an answer, we have always somehow snuck it in as a false assumption or created it by effectively dividing by zero.

Hypersphere cosmology by contrast posits a universe both finite and unbounded in space and in time. Nothing can escape it as the space curves back in on itself under its own gravity, as does time which just goes round in an immensely vast circle, but exactly the same things do not have to happen again. It has no outside because no space and time exists beyond it. Everything within it becomes recycled over billions of years, its entropy remains constant, on the very large scale it always looks roughly the same.

Neither model says 'where' the universe came from. Hypersphere Cosmology does not consider non-existence as somehow more fundamental than existence.

Philosophically I find HC the more satisfying of the two theories. This planet will not remain habitable forever in either model, but in HC the universe will effectively persist forever, so if we become smart enough, we can as a species also last forever. Thus, we should grow out of all those ghastly monotheist ideas about the end of days and the final judgement, take care of this planet as responsibly as we can, and start thinking about starships.

Plus, a universe which has effectively had unlimited time for life and intelligence to develop within it, will almost certainly contain some intelligences far in advance of us. I find that prospect fascinating.

Interview 12

348) What are your thoughts on simulation theory?

The idea that we may inhabit a computer simulation rather than the reality we perceive comes as the latest in a very long line of philosophical speculations about the possibly illusory nature of the reality we experience.

My English master at school once said, 'you cannot refute the proposition that the entire universe, complete with all our memories and records of it, was created five minutes ago'.

Well maybe you cannot refute it, but perhaps you can discard it on the grounds that the idea has no use or consequence, that we can neither confirm nor falsify it, and that it also implies something far more complicated than what it seeks to explain – a creation agency of some kind.

Since we have come to realize that we can probably, in principle, describe anything in the universe in terms of mathematics and information, some scientists have got a bit carried away with the idea that the entire universe consists fundamentally of nothing but mathematics and information. However, they fail to account for the observation that information only ever appears in 'embodied' form to us. We have never observed raw disembodied information, some medium always seems to carry it.

In previous eras some thinkers concluded that the universe consisted only of thoughts, and before that; that it consisted entirely of earth, air, fire, and water, or some other set of abstractions. In this they all fell for variants of the Neo-Platonic fallacy – the idea that all substance merely consists of congealed 'spirit', or 'essence', or 'thought' – in other words, that mind creates matter, or at least that something immaterial underlies matter.

> This idea seems to creep back in with quantum physics: "Everything we call real is made of things that cannot be regarded as real."– **Neils Bohr**

> The atoms or elementary particles themselves are not real; they form a world of potentialities or possibilities rather than one of things or facts." – **Werner Heisenberg**

In contemporary quantum field theory, the world that we can perceive seems based on interacting fields that we cannot perceive directly. Or if you like – that matter merely consists of a kind of interference pattern where otherwise invisible and intangible waves overlap, and the invisible 'wave' part of reality has vastly greater complexity than the 'particle' part we observe. Personally, I prefer to regard waves and particles as the origins of each other, with neither as the more fundamental.

Some magicians have toyed with the idea of using Simulation Theory as a magical paradigm, but this only makes sense if you can 'hack' the simulation in some way, perhaps by doing strange things which cause the simulation to malfunction in your favour, or perhaps by invoking the programmers (as gods?) for favours.

The idea of a 'Perfect' simulation has only a religious function, and then only under the assumption that the programmers may respond to appeals or be prepared to 'reincarnate' you.

Simulation Theory, as a thought experiment, also opens, or reopens, some wider philosophical questions. My old friend Lionel Snell posed a paradox with the question 'Can you precisely specify any human ability that a computer could not acquire?'.

It seems that you cannot. If you can precisely specify that ability then you could, in principle, design or program a machine to have it.

We do not usually want our machines to exhibit 'free will' but if we programmed machines to devise responses based on logic, quantumly random choices between some of the logically derived most optional responses, and a few other internal feedback loops with some random input, then we could make a machine yield outputs with any required degree of free will. Programming such a machine to vehemently assert that it did have free will (most of the time), and to make subsequent decisions usually biased by its previous ones, would then create behaviour indistinguishable from our own.

'Consciousness' really only means that some subsystems monitor other subsystems.

Machine learning programs have started to evolve in this direction. Watch out!

349) What about multiple universes?

The multiple universe idea comes from two different directions, the cosmological and the quantum. Cosmological multiple universes can result from a liberal interpretation of the questionable cosmological idea of inflation. Cosmologists sometimes invoke the idea of cosmic inflation to explain the large-scale uniformity of the observed universe. In the inflationary scenario, this universe begins with an extraordinary expansion of space itself at a speed enormously in excess of the speed of light. A secondary 'big bang' expansion of mass and energy into the almost instantly created vast space then occurs. Some cosmologists speculate further that inflation may have created a space almost infinitely larger than the observed universe and that many other big bangs caused by expansions of mass and energy may have occurred there, giving rise to a vast number of other completely unobservable universes at fantastically vast distances from ours.

Sometimes this 'multiverse' hypothesis becomes invoked to explain our current inability to say why the various physical constants of the observable universe have the values we observe and not others. In some of these universes within the multiverse, the speed of light, the gravitational constant, Planck's constant, and a whole host of other constants may have different values. Atoms and stars and planets may not even exist in some universes according to this hypothesis. In ours they only exist by chance because a random selection of physical constants has allowed it.

This all sounds rather hyper-speculative and barely scientific at all, and unless perhaps 'wormholes' can form in space and time to allow something to pass between such universes in a cosmic multiverse, then the whole idea remains useless and meaningless because it has no consequences whatsoever and we can never confirm nor falsify it.

Some theorists have abstracted the idea of multiple universes from quantum considerations and this continues to attract believers in the idea of 'parallel universes'.

Our current quantum theory can only specify a range of possible pasts, presents, and futures for the observable states of the quanta which seem to underly reality. The quanta of matter and energy seem to behave like particles when they interact or 'land' or when we observe them, but when they 'fly' between such events they seem to fly as waves of probability. They seem to spend nearly all of 'their' time, or time as we measure it, in flight, in states that we can only represent as a mixture of probabilities, but this mysteriously collapses into a definite state on the rare occasions when they actually do something by interacting with each other or with our measuring devices.

When they do something, what they actually do often seems to show some influence from all the other things they could have done in their past and perhaps all they could have done in their future as well.

The parallel universe or multiple universe hypothesis seeks to resolve the mysterious transition between probabilistic wavelike and definite particle like behavior by stating that quanta always do everything that their wavelike nature permits but we only see them do one of those things.

I some versions of this idea, every transition of every quantum from wavelike to particle like behavior creates alternative physical realities in which that quantum can execute every one of its possible behaviors. This would imply that the greater universe must contain at least ten followed by five hundred zeros worth of sub-universes, with the number rising continually.

One has to wonder just what the greater universe makes all these alternative universes out of, and where it puts them so that we cannot see them.

In milder versions of the hypothesis the alternative realities all exist all around us but they rapidly interfere with each other and decohere back into the single observed particle reality. This doesn't seem much different from a wave-particle model in which quanta fleetingly phenomenising as multiple waves rapidly revert back into single particles.

Perhaps the tea you chose instead of coffee in this universe may have an infinitesimal hint of coffee about it - if it proved a tough choice. But I cannot believe that the choice creates two separate universes in which you drank different beverages and subsequently led slightly different lives.

I suspect that much of the quantum weirdness arises because quanta do things in more than the three dimensions of space and the one of time that we commonly acknowledge. We do not yet see the whole picture, but I do not think that such additional dimensions contain multiple particle-based versions of the universe. Rather I suspect that whilst particles appear as four-dimensional events, their associated waves operate in six dimensions and we cannot perceive or easily intuit the other two because they consist of extra time dimensions.

Contra-Factual Imagination – the ability to conceive of what does not yet exist and of what may not even exist at all, seems a large part of what makes us human. No other terrestrial animal can do this to any great extent, but we have imagined everything from the great pyramid, to gods, and to spacecraft, and made some of them real.

Every novel and every play ever written, and every mythological story we have ever told each other, counts as an alternative universe in some sense.

I particularly like one definition of Magic which runs as - 'The Science and Art of using imaginary phenomena to create real effects'

350) What do we know or think we know about other dimensions? Is this idea much more useful than the multiverse?

Yes indeed, adding extra dimensions to the observed universe to model and explain the peculiar behaviors that scientists and magicians observe, seems far simpler and more parsimonious than trying to explain them by concocting entire extra universes.

The popular imagination often seems to regard 'other dimensions' as meaning 'extra universes' where peculiar things happen that can perhaps occasionally have effects on our observed universe.

I prefer to consider a rather stricter interpretation of the idea of 'dimensions' in terms of the number of coordinates or degrees of freedom we need to describe our observed reality.

Gravity in some senses acts as an extra dimension. Consider the surface of the earth, it looks flat over short distances but very large triangles drawn on its surface have angles that add up to more than 180 degrees, and very big circles encompass more area than expected for a flat surface. Gravity curves space and it also curves time, time slows near massive objects. We can regard this curvature, which we observe as an acceleration, as an extra dimension. The gravity of the entire universe almost certainly makes it a closed structure and hence a hypersphere best described as a structure with four spatial dimensions with its curvature representing its fourth dimension.

Quantum phenomena seem to exhibit degrees of freedom that we cannot well describe even by using a spacetime of four spatial dimensions and one time dimension, and even if we allow phenomena to work both forwards and backwards across one dimensional curved time.

This has led some theorists to hypothesize extra hidden dimensions in which the quanta exercise their peculiar extra degrees of freedoms, like their ability to seemingly exist in several different states or places or times at once. Most of these ideas depend on one, or six, or a rather large number of invisibly small extra spatial dimensions. Such ideas have so far demonstrated only limited descriptive power and almost no predictive power. 'String' and 'Brane' theories with many spatial dimensions have yet to bear much fruit.

Hypotheses of extra time dimensions remain in their infancy, but I find them far more interesting. Extra full size time dimensions could easily pass unnoticed, we cannot even 'see' one dimensional time, we merely infer it from memory and expectation. Both quantum and magical experiments indicate that both the future and the past have a multiple rather than a singular nature and that they can both influence the present moment of observation. This of course suggests that we live in a universe that runs on probability rather than strict causality, or that causes lie hidden in sideways times. On general grounds of symmetry and to account for the multiple trialities of the particle physics realm, I suspect that time has the same (curved) threefold dimensionality as space.

351) What might be some characteristics of three-dimensional time?

I suspect that having curved three-dimensional time would give rise to a universe having precisely the characteristics that we observe in this universe!

Quanta (fundamental particles) can exhibit four degrees of freedom called spin, electric charge, nuclear charge, and generation or flavor. Spin uses up two spatial degrees of freedom; generation uses up the third spatial dimension and the curvature; the three temporal degrees of freedom plus the temporal curvature can account for the electric and nuclear charges. That summarizes, with ridiculous brevity, the hypothesis that I currently pursue with several collaborators.

All the equations of physics (with the possible exception of the second law of thermodynamics) work equally well forwards or backwards in time. Anti-particles such as positrons correspond to electrons with positive charges that travel backwards in time. The existence of quanta with electric and nuclear anti-charges suggest that these characteristics arise from additional temporal degrees of freedom, rather than spatial ones.

On the more everyday scale we seem to inhabit a universe that runs on probability rather than strict causality. Three-dimensional time implies that the many different futures remain possible. It also implies that many different pasts remain possible, even though we commonly assume that only a singular past could have led to the observed moment of the present.

Some quantum physics experiments and some magical experiments do seem to show some kind of interaction between the present and a number of possible pasts and futures, and perhaps also with alternative presents superposed in sideways time.

A spatial analogy can perhaps show why time appears only one-dimensional to us. Imagine a blindfolded person attempting to walk forward (or backward) in a landscape without obstacles. Better still, imagine a blindfolded scuba diver swimming forward in a vast expanse of water with no sensitivity to depth from pressure. Both will assume that they move in a straight line, the walker could remain unaware of the full two-dimensional nature of the landscape, the diver could remain unaware of the full three-dimensional nature of the ocean.

We cannot 'see' time. In practice both the walker and the diver could create or sense accelerations from sideways motions, in temporal terms these correspond to choices and probabilities, in three-dimensional time we have choices rather than a single fixed future, and sometimes improbable events occur.

352) But that model of time doesn't give us new avenues of exploration where time travel or anything more out there is concerned?

In this model only the present moment has a particle like physical reality, the pasts and the futures of the present moment have only a wavelike and probabilistic existence, so no solid reality exists there for us to visit. We can of course simply wait and see what particulate reality the universe rearranges itself into in the future, but no mechanism exists for recreating any past in particulate form – at least not on the human scale macroscopic level.

Relativity does at least offer the opportunity to reduce the waiting time for the future. If you loiter in a strong gravity field or under a strong acceleration or at a very high velocity then you will experience a slowing down of your own time and find that the rest of the universe has done more than you have. This effectively provides time travel to the future, but it does not recreate a past that you can physically visit.

This galaxy contains many billions of stars, the universe contains many billions of galaxies. The majority of stars seem to have planets, a few of them may support life and a few of those may support intelligent life. The universe thus probably contains many billions of different intelligent life forms. That sounds like plenty out there to me.

In principle, relativity could allow us to reach a star system a hundred light years away in what seems like a few days, although if we made a return journey, we would arrive home in what also seemed like a few days, to find that two hundred years had passed.

Quantum entanglement means that where particles have had contact, or where particles have exchanged radiation, the particles seem to have some sort of instantaneous connection over any distance. This indicates to me that waves can travel backwards in time at 'reverse lightspeed'. When my eye catches a photon from a star a hundred light years away it instantly modifies the way in which the star emitted it a hundred years ago. We may one day work out how to use this effect to communicate with aliens on distant worlds. As both we and the aliens would effectively lie at exactly the same 'distance' in each other's past, we could in principle communicate in real time.

353) Why do you think humor developed?

If we take 'humors' to mean emotional states of mind in the most general sense, then humor in the specific sense of laughter seems to have developed to moderate our humors in the more general sense.

Emotions seem to have evolved as quick decision-making mechanisms; avoid pain, seek pleasure, fight, or flee, try to dominate, or submit, press sexual advances, or withdraw, and so on. Of course, 'thinking' can add a layer of complexity and moderation on top of this, and emotional behaviors can become ingrained and dysfunctional.

Many of the more complex social animals such as apes other than us, dogs, rats, and dolphins all seem to exhibit something that looks like laughter, and they mostly seem to do so when play-fighting or tickling, presumably to defuse the impulses to full scale dominance behavior, serious violence, or immediate sexual activity.

As we over-brained apes have developed ever more complex social behaviors and relationships and self-identities and an awareness of death and ever more complex philosophies, we have developed an increasingly sophisticated sense of humor to take some of the sting out of these experiences and to moderate our behavior.

All humor experiences seem to arise in response to some kind of dissonance, perhaps between what we perceive or believe or expect and what we experience or what we or others expect to perceive or believe or feel expected to believe. Thus, suddenly understanding an equation which shatters preconceptions about the structure of the universe can elicit as much laughter as seeing someone break wind whilst pretending to high dignity.

Let us face it, life itself seems a crazy wasteful extravagance. All living things will die. Almost nothing that we strive for will matter one whit a hundred years hence, the absurd human soap opera full of fears and desires, love and war, sex and death, personal identity and even the attempt to transcend it, will mostly come to naught in the grand scheme of things.

In a universe and a human condition that does not favor sanity, laughter helps preserve it.

In Chaos Magic we use laughter as a 'banishing' after performing some conjurations, particularly after enchantments and invocations. This has the effect of inhibiting the discursive thinking which can undo the intent of the conjuration, and it also helps to reduce the megalomania/self-doubt which the practice of magic can engender. Laughter, even if at first forced, becomes natural with practice. In Chaos Magic we give the conjuration everything we have got and then laugh it off and await the results with equanimity.

354) What about other dimensions of space? Or dimensions we don't perceive? Can pretty much all phenomena that's been attributed to the multiverse or whatever be explained by different dimensions?

The word 'dimension' can have a variety of meanings but they all revolve around the idea of degrees of freedom. Consider a conventional chessboard, the pieces have various freedoms to move within just a two-dimensional plane and we only need two coordinates to specify the position of a piece. We can also make three-dimensional chess games, for example a game set in an 8 x 8 x 8 cube in which the pieces have various freedoms to move in the up and down directions as well. In this case it takes three coordinates to specify the position of a piece, and the board then has 512 cubic cells and a new type of possible move involving transiting through the vertices of cubes as well as through the faces or edges.

We can add another abstract dimension represented by say color and give the pieces various freedoms to move through a spectrum of say eight colors with say a rule that a cubic cell has the color of the piece which occupies it and that only pieces of the same color can attack each other. By now we have an almost unplayable complex game in which the pieces have four degrees of freedom, but we can do worse still:

We can also alter the topology of the play area and consider that opposite sides of the flat board connect so that a piece can move off one side and immediately back on to the opposite side. In the case of three-dimensional chess, opposite sides of the cube connect directly to each other and long-range pieces like queens, bishops, and castles can move in great circles. This creates hypercubic chess in curved space, although it remains three dimensional, the curvature adds another degree of freedom.

The record of a game of chess provides its temporal dimension, it shows what happened over time. It does not show the moves not taken. However, we could in principle have multiple timelines. A player unable to decide between two moves could elect to make both of them, perhaps by replacing a single piece by two smaller shadow pieces in different positions with the provisos that whenever one of the shadow pieces interacts with an opposing piece, the other shadow piece becomes removed, and a player moving a shadow piece must also move its corresponding other half during a move.

Strangely, afterwards, we could write a single timeline to describe the whole game that does not include the shadow pieces that became removed! However, an expert would probably notice something rather odd about some of the moves chosen when looking at the record afterwards. I think that such a Quantum-Chess analogy can in principle explain many of the mysteries of quantum mechanics and magic. What may have happened, and what could have happened, can affect what does seem, to us, to have happened.

Of course, when occultists refer to other dimensions, they often mean something analogous to other chessboards hidden out of sight where gods and demons and angels, spirits and ancestors play games that can sometimes have effects on the games we play here. Such hidden chessboards exist inside every human head.

As above so below. In both the physical and mental dimensions, imaginary phenomena can have very real effects. We all know what 'imaginary' means in the mental dimensions, in the physical dimensions 'imaginary' means phenomena which can remain unobservable except as probabilities.

355) Do you think humor is evolving?

The forms of humor seem to have remained constant for all of recorded history but the subject matter of humor seems to evolve more quickly now. The classical Greeks enjoyed all the forms of humor that we appreciate today from slapstick physical humor through ridicule to sophisticated political and social satire. Throughout history we have punished each other for laughing at the 'wrong' things, and in these days of political correctness, thought-speak, cancel-culture and no-platforming, the list of unacceptable topics has grown enormously. I loathe that word 'unacceptable'. Unacceptable to whom? Beware of any person or group that cannot take a joke cracked at their expense and learn something from it. I do wonder if many contemporary comedians have fallen back on mere self-mockery and self-ridicule because anything else may give offense to someone somewhere.

Too many people now seem to regard their identity as something sacrosanct and as an end in itself, rather than just a role or a game that they play.

356) Or in the sense of seeing what's possible and taking it further, do you think magic can evolve?

I have seen magic evolve a lot during my own career and it has evolved in the direction that I predicted from my Psycho-historic model of the Aeons.

Magic has increasingly dissociated itself from the religious ideas and motivations that it used to employ to give itself a vocabulary and a philosophy. Today's magicians want immanence rather than transcendence. They want results on the material plane and they want to understand their own minds. Deities have become recognized as elaborate metaphors for abilities we can invoke. Spirits have become recognized as thought forms that we can evoke to create effects. All the procedures of enchantment and divination have become recognized as tricks that magicians play on themselves to liberate their parapsychological abilities.

In short, magic has moved away from religion and closer to science.

Thelema or Crowleyanity and Neo-Paganism in general represented the last major flowerings of religious flavored magic in the western tradition for some people. Yet many others took a far more agnostic or psychological view of the gods and goddesses and 'spirits' of these systems. Chaos Magic adopts the working hypothesis that humans create deities and spirits for their own purposes, rather than the other way around. It does however take a liberal view of confirmation bias when exploring experimental beliefs.

The Psycho-historic model the Aeons does predict both a comeback for transcendental religious type ideas and a decline in scientific-materialist thinking in the distant future so we may eventually see new forms of magical thinking aligned with new transcendental ideas Whether this will take place in the ruins of a post-apocalyptic future or whether it will take place in a hyper-tech future in which nobody really understands the technology they use anymore, remains to be seen.

357) Do you think there will be a point at which our models are as complex as our nervous system will allow them to be?

Human knowledge in even just the physical sciences now far exceeds what any single human could possibly absorb, and the Renaissance polymath's dream of knowing everything has long gone. There seems an enormous amount that we still do not know about the universe and quite probably a lot that we do not yet realize that we do not know, the unknown unknowns. The perennial question of epistemology versus ontology remains with us, how well do our descriptions of reality match reality? We may find that our descriptions of the quanta and the cosmos become entirely mathematical and devoid of any possible useful visual or verbal analogies. We could end up with models that fit the most sophisticated observations but which we can no longer understand in any meaningful way. This might lead to the development of new mythologies to fill the gap.

358) Do you have a favorite theory on how language evolved?

The development of human language seems to have depended on four major factors. Firstly, humans live as social animals. Solitary creatures like many reptiles which do not care for their young or cooperate in defense or finding food would gain no advantage from it. Secondly, humans have bodies that can extensively modify their own environment by shifting materials around and making tools, shelters, and weapons. Thirdly and fourthly humans developed the vocal apparatus and the neurological circuits to make and understand sounds. These last two depend critically upon each other and they must have developed in tandem. Once they started to develop, they would have conferred such huge survival advantages that they became heavily selected for.

Perhaps fifthly we should add hyper-suggestibility. Humans seem both blessed and cursed by an extraordinary ability to tell each other what to do and to imitate each other. Children everywhere quickly acquire their native tongue and most of the amazing and appalling aspects of their host culture.

Of course, all social animals have some rudimentary means of communicating with each other but they lack the complicated lifestyles that human languages have evolved to support and enable.

The development of particular human languages remains a topic of intense debate. Some show fairly obvious signs of common origins, others mysteriously do not. Languages have evolved a great deal in the course of a few thousand years of recorded (written) history and they presumably evolved a great deal more during the several hundred thousand years of pre-recorded speech, so we will never discover the earliest form or forms of language. No single word stands for the same single concept in all known human languages.

The extent to which a language controls what its speakers can and cannot think about remains a fascinating and controversial topic. Yet rather than engaging with arguments about cultural and national stereotypes, I would prefer to look at the language of mathematics and the concept of 'being' which has a representation in most languages.

All languages enable the formation of contra-factual concepts; thoughts about what might happen and what might have happened, thoughts about things that do not exist to our direct sense perception, and all our imaginations, hopes, fears, lies, and guesses.

Mathematics provides a language that most humans can agree on when it comes to simple arithmetic. Even negative numbers make sense when equated with the idea of a lack of something or a debt owed or time passed. We can unequivocally translate between say numbers in Indo-Greek base ten mathematics and Babylonian base sixty mathematics and the cumbersome Roman numerals. All these numbers relate to amounts of distance, or things, or time. Yet the so called 'imaginary' numbers which have proved essential in describing certain electrical and quantum phenomena have no corresponding meanings in any verbal language, so mathematicians have evolved a language in which they can think what others cannot.

The concept of Being, the verb 'to be' and the concept of 'is' have representations in most languages, yet these represent imaginary rather than sensed qualities. We never experience something just 'being' everything always does something. Nothing 'is' anything else, it merely resembles something else in its behavior. The semanticist Korzybski recommended that we drop the use of this imaginary concept in language and adopt V-prime language instead. Vernacular prime avoids all tenses of the verb 'to be' on the grounds that the ascription of 'being' or 'is-ness' to anyone or anything merely leads to a shorthand of dubious assertions, prejudices, and false premises. The concept does not exist in mathematics. The 'equals' sign more modestly denotes that one thing has the same numerical value as another when measured in specified units.

Some philosophers, notably Wittgenstein, have opined that philosophy fundamentally consists of arguments about the meanings of words, and indeed philosophy or philo-sophia originally meant word-wisdom, although today as we struggle to understand the quantum fundamentals of reality, it should also include arguments about the meaning of the weird mathematics that describe it.

359) Do you think it's possible to create or discover a technology that will revolutionize our experience as much as language did?

Yes, and I will venture a guess as to what could do that. Most people tend to think in images at least as much if not more than they think in words, yet capturing mental images remains difficult for us, we usually have to try and reduce them to words or try to paint or draw them.

What if we could develop some sort of direct Brain-To-Screen imaging system?

Imagine putting on some sort of a helmet and sitting down in front of a screen and projecting memories, imaginations, and dreams straight onto it, or onto someone else's screen.

The possible uses and abuses of such a technology seem endless. Would we ever look at ourselves or anyone else in the same way again having peeked inside?

Interview 13

360) Does hypersphere cosmology or three-dimensional time have any new implications for the way we think about black holes?

In Hypersphere Cosmology black holes all contain hyperspheres rather than spacetime singularities. Black holes still have the usual Schwarzschild radius when viewed from the outside but within that radius the geometry of the closed region of spacetime becomes hyperspherical and governed by the Godel metric, which means that it also spins with any point moving at lightspeed with respect to its antipode point.

In this model, a hypersphere of a given mass will have a constant size and the mass within it will remain uniformly distributed and churning around with no real center.

I strongly suspect that the entire universe, individual black holes within the universe, and all quanta in particle mode, all consist of hyperspheres.

This hypothesis leads to potentially testable consequences in all domains:

The hypersphere of the entire universe never changes size, the cosmological redshift arises from a gravitational space and time dilation effect, the apparent accelerating expansion arises from the gravitational lensing of space. The big bang did not occur. The angular velocity of the galaxies around the universe about multiple axes remains potentially measurable. The spacetime geometry inside of a hypersphere the size of the universe will appear very close to flat Euclidian space and the overall average density will appear very close to what we observe. However, it's very small positive curvature will create a small omnidirectional deceleration which will give rise to anomalous galactic rotation curves (which some physicists have misattributed to a mysterious dark matter) and also a resistance to linear motion which we have already measured in the Pioneer Anomaly.

Individual black holes within the universe will experience a spin up effect from the entire universe which will tend to destabilize them in proportion to their size and cause them to slough off mass and energy, in addition to that lost through Hawking Radiation. Thus, black holes will tend to decay over time, recycling matter back into space. Note that all observed black holes appear to emit radiation.

From the outside, black holes/hyperspheres exhibit only three properties, mass, charge, and spin, and these three properties also define fundamental quanta like electrons. At the quantum scale, the effects of three-dimensional time become quite large compared to space and they provide the extra degrees of freedom that the quanta have, such as both nuclear and electric charge and wavelike behavior. This part of the hypothesis remains under development.

So, within the hypersphere cosmology model the universe does not have a cataclysmic beginning or ending, on the inside it effectively remains without boundary in space or time although it will have spatial and temporal horizons at thirteen billion light years and thirteen billion years respectively. Black holes within the universe do not represent a final degenerate state of matter from which it cannot escape, they evaporate over time.

361) Are you an optimist or a pessimist?

Well, as the old quip goes, the optimist considers this to be the best of all possible worlds - and the pessimist has the same opinion. Optimists describe a glass as half full, pessimists describe it as half empty. Now approaching seventy years of age, I choose to regard my glass as seventy percent full and thirty percent empty. I seem to have accumulated miraculously little damage to mind and body so far and I might just have enough of each left to make it worth surviving to a hundred. There remains much that I want to fill my glass with, and I feel optimistic about my chances of achieving some of it. Life comes at the price of Death, a good bargain in my view. Without death, life would lack motivation and become stagnant.

Yet I do sense and feel a rising current of pessimism about the future of civilisation as we know it. As a scientist I can see that on current trends the maths does not add up to a civilised future for humanity. The ever rising production and consumption, the ever rising population, the ever rising pollution and environmental degradation cannot continue much longer on this finite planet. I may well live long enough to see very radical changes and I feel concern for my descendants' futures.

If as a species we do not somehow collectively choose to hugely downsize our lifestyle then that downsizing will become forced upon us by a gradual or acute decline of industrial civilisation and all that comes with that, including the usual Horsemen of the Apocalypse – War, Famine, Pestilence, and Death.

I do not feel particularly optimistic that we can develop a whole suite of technical fixes in a hurry and carry on economic and population growth as usual. Since the end of WW2, 'progress' has gone global and exponential but a number of analysts reckon that after the mid 1970's it ceased to deliver any greater general life satisfaction in the developed world, and since then, 'progress' has had a gradually worsening effect on general life satisfaction and wellbeing. I would happily embrace a return to the material levels and attitudes of the 1970's, a decade I enjoyed very much, but I would feel less enthusiastic about civilisation returning to any previous period.

362) What are the most meaningful statements we can make about human nature?

We must never forget our glorious simian heritage.

The Irish Elk evolved antlers three times bigger than in other species of Elk. It did not prove a viable evolutionary strategy and it went extinct. One species of Ape has evolved a brain three times bigger than in other species of Ape. It remains to be seen whether this strategy will prove viable.

Expect the worst, hope for the best, that way you only get pleasant surprises.

Human nature seems to exhibit such plasticity and mutability that some wonder if it even exists. As the most suggestible of all known species, we can easily persuade each other to believe or do almost anything. People tend to do whatever they think a situation demands of them and to adjust their beliefs and feelings accordingly. We have a far greater propensity to believe what we do, rather than to do what we believe. Biography tends to define Philosophy.

Famous people in history often showed such complex characteristics that all pronouncements about their supposed 'nature' seem inconsistent. Winston Churchill joined in a battle that he didn't have to at Omdurman 1898, and killed several opponents at close quarters, yet he painted sensitive and exquisite watercolours and lived as a devoted family man. Unlike Adolf Hitler who regarded it as cruel, Churchill enjoyed eating plenty of meat.

I prefer a Pagan Polytheist view of human nature over a Monotheist or Atheist Individualistic view. Humans naturally have a multiple nature, we can think and do mutually incompatible things, we can become completely different people in different situations, we have no fixed nature, we have worlds within us, we consist of a Pantheon of selves.

The classical pagans realised and accepted this and gave homage to all their abilities and proclivities in the form of deities, Mars for war, Venus for love, Mercury for intellect, Jupiter for prosperity and power, and so on. Monotheism tried to overturn this with the idea of a singular nature or self or personality to which we must all adhere, rejecting everything else. The mismatch between this idea and reality fills the monotheist mind with demons and initiates civil war.

Post-Monotheist Atheistic-Individualism does little better. The quest for authenticity and truth to oneself will always fail because we consist of a lot of selves with conflicting agendas.

To live a full life and to avoid war in heaven, our multiple selves need to allow each other to express themselves in any way that does permanently affect our other selves' expression.

363) Is there a secret of life?

Yes, and Erwin Schrodinger (who also developed the Schrodinger wave equation) correctly deduced its nature in 1944. The secret of life lies in an Aperiodic Crystal. In 1953 Watson and Crick (and Rosalind Franklin) precisely identified this Aperiodic Crystal as DNA. Life consists of DNA replicating itself in increasingly complex and baroque ways.

From this alone we can deduce the secrets of the Human Condition.

For several billion years of its terrestrial life, DNA did not bother to evolve forms with nervous systems capable of complicated abstract thought and imagination. It tried just about everything else from bacteria to jungles to hundred ton Whales. In us it has conducted an experiment, and we have now become conscious of it.

Our DNA programs us to eat, sleep, reproduce, and die. Our sleep and death remain relatively simple, but large brains have led to our eating and reproducing behaviours becoming fantastically complicated. The desire to eat has led us to master fire and shelter and clothing to find food resources in difficult climates. It has led us to make weapons to kill game and to fight with competitors for food or mates. It has led us to farm and to create social stratifications and civilisations.

Our desire to reproduce has similarly flowered into highly complex expressions, many of them only distantly connected to making more DNA. We pursue sexual activity for its own sake, and sexual partners for reasons other than their reproductive potential alone. We take exceptional and prolonged care of our offspring and concern ourselves with matters of curiosity and legacy and perhaps the survival and vitality of the tribe or even of the whole of humanity. All our art and culture, religion, science, and magic has its roots in a sublimated urge to reproduce ourselves and our DNA.

DNA fights with itself to make itself stronger through the bloody and profligately wasteful process of evolution. It continues this fight with itself through us, and within us.

The secret of life lies in the recognition that life does and should consists of perpetual struggle, and for an over-brained Ape only an examined life seems worth living to me, although many do seem to subsist on an unexamined life.

Pleasures evaporate quickly. The greatest satisfactions in life arise from achievement and in becoming of interest or service to others.

364) Self-image?

'O wad some Power the giftie gie us

To see oursels as ithers see us!'

- Robert Burns, To a Louse, 1785.

Recordings of my voice always sound strange to me, photographs of me don't look to me like what I see in a mirror. Perhaps my self-image does not match what others see in me. From a young age I have had a fascination with ideas, an aversion to authority and received wisdom, a reflex to question everything, and a strong desire to imagine and innovate alternatives. I suspect that some people regard me as eccentric or perverse for its own sake, but I see myself as making a continual effort to tone it down in public.

I have had a lifelong fascination with board games for example, but I spend far more time designing and building them than playing them. I cannot open a box game without soon starting to redesign and enlarge its structure and concepts.

I have an aversion to following rules and instructions and I do not consider myself a team player, I have no interest in any team unless I command it. I don't regard this as arrogance, I simply prefer to make my own mistakes rather than someone else's. Yet I probably do have adequate reserves of arrogance and self-regard – very few people attract my unreserved respect – I have few heroes.

The Big Five Personality Traits – OCEAN - openness, conscientiousness, extraversion, agreeableness, and neuroticism provide only a rough sketch of anyone that tends to remain rather circumstance dependent, but I would score myself out of ten on each of those traits as follows:

Openness to experience (inventive/curious vs. consistent/cautious). Nine out of Ten for interest in novel ideas or experiences, but I will just as readily reject anything new if I don't like it. 'New' or 'Modern' frequently seem to mean worse than what we had before, particularly in the fields of modern art and architecture, popular music, and moral fashions.

Conscientiousness (efficient/organized vs. extravagant/careless). Three out of Ten. I save my limited conscientiousness for things which seriously matter to me; with everything else, 'good enough seems better than not at all'. Imperfection does not bother me in most matters, dust, extravagant clutter, untidiness, and careless dress characterise my lifestyle and I find them oddly reassuring.

Extraversion (outgoing/energetic vs. solitary/reserved). Four out of Ten. I have no great enthusiasm for public speaking or loud parties, I find such things exhausting, whereas my own company or that of a book or a small number of friends usually proves invigorating. I don't like loud people.

Agreeableness (friendly/compassionate vs. critical/rational) Eight out of Ten. Despite my low expectations of human nature, I always hope for the best and I have perhaps expended too much courtesy, politeness, and niceness on people who weren't worth it, but it has paid off in terms of those who proved themselves worth it. I have sometimes played the game of agreeing with persons whose opinions I find disagreeable and feeding those opinions back to then in amplified form, but I also enjoy intense but civil argument.

Neuroticism (sensitive/nervous vs. resilient/confident) Two out of Ten. I have a few not entirely rational attitudes. I avoid offal, shellfish, and nitrate preserved meats as potentially harmful despite a lack of medical consensus on these matters. I avoid standing anywhere close to working microwave ovens. I don't like unenclosed heights because of an irrational subconscious desire to throw myself from them. On the other hand, I feel generally impervious to any form of criticism. Criticism merely reveals the limitations of my critics. My fortress of arrogance has awesome resilience beneath its cloak of humility.

365) Can you explain Hypersphere Cosmology?

Yes, it all depends on: Perhaps I should clarify that in an appendix, a mere sixteen pages should do it.

See Appendix – The Occultaris part 1, Hypersphere Cosmology.

It's been a pleasure talking with you Ian.

It's been an honour and a pleasure talking with you too Pete, maybe we can do it again in ten years!

Appendix, The Occultaris part 1

Hypersphere Cosmology
What surrounds us?

For thousands of years humans could only speculate about what if anything existed beyond the sun, the moon and the planets and the apparently fixed stars visible in a surrounding sphere. Usually, they hypothesised that gods and goddesses much like themselves inhabited the celestial sphere and any realms beyond that. Only in the twentieth century did it become apparent, from new powerful telescopes, that our sun forms part of a galaxy of several hundred billion stars and that it seems a rather average specimen of a star. Perhaps even more astonishingly the observable universe contains at least several hundred billion entire galaxies, but no observable deities.

However, twentieth century astronomy did offer the religiously inclined a possible consolation because although the universe stood revealed as more awesomely vast than any religious revelation had ever suggested, it did at least appear to have begun with a creation event, the so-called Big Bang, and because this creation event seems inexplicable *in principle,* maybe a god did it.

This chapter will make the case that no such event occurred and that all the apparent evidence for it quite readily submits to an alternative interpretation and an alternative model of the universe. It will show how the Big Bang theory developed in response to the order in which astronomers and theorists made and incorporated new observations and made new assumptions. It will show that an alternative model of the universe can also accommodate all the known observations without invoking the dubious ideas of spacetime singularities, cosmic inflation, cosmic expansion, dark matter, and dark energy.

This alternative model, Hypersphere Cosmology (HC), describes the universe as Finite and Unbounded in both space and time. The universe's own gravity closes it in on itself so that sufficiently long voyages will return travellers to their (by then unrecognisable) points of origin after many billions of years. Space exists only within the universe. We have no reason to suppose that space or time exist outside of it. Time also has a finite but unbounded nature like the surface of a sphere. It has no beginning or ending but no two events will appear to have a separation in time of more than about thirteen billion years. Such a cyclic form of time does not mean endless repeats of the same events, the observable point of the present moves randomly around a temporal hypersphere rather than around a simple circle.

The Hyperspherical universe has no beginning or ending in spacetime and no spacetime exists 'outside' of it. We have no reason to imagine non-existence as somehow more fundamental than existence. Although we can create privative concepts like 'non-existence' or 'space as absolute void' or 'time as empty interval', such phenomena have no reality. Nothing comes from nothing. All things arise from rearrangements of other things and will eventually become rearranged. Hypersphere Cosmology embodies a paradigm change in thinking from the western neo-monotheist linear view of time with a beginning and an end, to something more akin to the cyclic perspectives of some forms of Paganism and Buddhism.

This chapter will present Hypersphere Cosmology firstly in prose without mathematics and compare it to the Big Bang model. The prose will invoke a certain amount of visualisation of unusual geometries. The chapter will end with a presentation of the mathematics and quantitative data that support the case for the universe having a hyperspherical geometry. Although mathematics, particularly algebra, proved crucial to the elucidation of the hypersphere model, it remains perfectly possible for the mathematically disinclined to understand it in a purely qualitative manner. Personally, I regard the algebra as the poetry.

Bold numbers in the text **(N)** refer to the equations in part 2.

The story of intergalactic scale cosmology begins with Einstein's theory of General Relativity published in 1915 which provided a geometric theory of gravitation. It describes mass and energy as curvatures in spacetime that extend across spacetime to create the effect we call gravity. It predicts 'black holes' – areas of spacetime that have become so highly curved that the resulting gravity prevents anything including light from ever escaping. Einstein thought that the entire universe probably existed inside a black hole, closed by its own spacetime curvature, it appeared to have about the right amount of mass and energy density in it for closure despite its immense size. This would make the universe a hypersphere, an inescapable bubble of spacetime with the odd property that really long journeys within it would return travellers to their points of departure.

However, Einstein had a problem with his hyperspherical universe, there seemed nothing to stop the spacetime curvature or gravity of all the galaxies pulling them all towards each other with increasing speed and causing a catastrophic implosion of the entire universe. Reluctantly he added a fudge factor which he dignified with the name of the Cosmological Constant. This added enough 'anti-gravity' of unknown origin on the cosmological scale to prevent collapse, simply on the basis that as the universe had not already collapsed then something like that had to exist.

This unsatisfactory situation persisted until the assumption of a basically static universe became questioned a decade or so later. Light from distant galaxies shows a 'redshift' that resembles the Doppler effect observed when a source of sound moves relative to an observer. For example, an emergency siren seems to emit a higher note from a vehicle hurtling towards an observer and then a lower note as it hurtles away from the them. The light from distant galaxies all seemed to show a shift towards the lower (red) end of the spectrum and the shift seemed to increase in proportion to how far away any galaxy lay in space. The redshift became interpreted as indicating the speed with which a galaxy hurtles away from us. The work of Friedman, Hubble, and Lemaitre culminated in a challenge to Einstein's view of a static universe. By around 1930 Einstein conceded the argument, retracted his cosmological constant, and accepted the expansion idea.

The theory of the expanding universe gradually became hailed as a supreme scientific achievement and a prime article of faith, amongst cosmologists as an incontrovertible fact. An expansion of the universe implies that in the distant past it must have occupied a much smaller volume. As theorists could not think of anything to limit just how small a size the universe could have expanded from, they concluded that it must have expanded from zero size, or at most from a size considerably smaller than a single atom. Despite that nobody could account for how such an initial condition could have arisen, cosmologists became convinced that long ago the entire mass and energy of at least several hundred billion galaxies erupted out of a space smaller than a single atom at what they came to call the Big Bang event. This theory seemed to receive confirmation when in 1965 astronomers observed a faint glow of Cosmic Microwave Background Radiation (LMBR) pervading all of space and concluded that this radiation came from the intense fireball of the early universe which had subsequently cooled as the universe expanded.

However, some dissented from the expansion theory. In 1949 Kurt Gödel, a close friend of Einstein and an extremely eccentric logician who remains famous mainly for his Incompleteness Theorem, developed a rotating universe solution to Einstein's equations. In this model the universe does not implode because it rotates, much in the same way as the earth remains in orbit around the sun and electrons stay in orbit around the nuclei of atoms. This solution became generally ignored and discarded because nobody could perceive any axis of rotation in the universe. Yet whilst a rotating sphere will have an obvious axis of rotation, a hypersphere will not as it has no centre or boundary, and it undergoes a special type of rotation of its own. Unfortunately, nobody at the time seems to have factored this in, because as this chapter will show, hyperspherical rotation not only prevents the universe from imploding, but it also prevents spacetime singularities from developing and it also accounts for the anomalous rotation of galaxies and galactic clusters.

Astronomers first noticed anomalies in the 1930s onwards and the work of Vera Rubin from the late 1960s onwards revealed that galaxies in general do not rotate as either Newton's or Einstein's equations predict. The outer regions of most galaxies rotate considerably faster than either theory predicts for objects of galactic size and mass. Instead of refining the theory to match the observations astronomers took the extraordinary step of changing reality to fit the theory. They decided that galaxies must contain a mysterious 'dark matter' which just seems to have mass and gravity but no other properties at all. To preserve their equations, they added about five times more of this peculiar invisible and otherwise undetectable stuff to the universe than all the observable 'ordinary' matter that makes up us, this planet, and all the stars and galaxies in the universe.

Then in the 1990s investigations into the precise relationship between the redshifts of galaxies (which we can measure easily and accurately) and their distances (which we can measure only with difficulty and with wider margins of error) culminated in the work of Perlmutter et al. This showed that redshift did not bear a linear relationship to galactic distances as cosmologists had assumed for decades. Galaxies with substantial redshifts seemed very much dimmer than expected and hence apparently much further away. Again, cosmologists took the strange step of altering reality to preserve their theories and missed another opportunity to sort out the theory. In this case they decided that the expansion of the universe must have started accelerating about halfway through its history. To account for this acceleration, they added about three times more 'dark energy' than the dark matter they had already added to the universe.

Dark matter could possibly explain why galaxies seemed to form so alarmingly quickly after a big bang if such an event had occurred, but the Dark Energy has to pop up out of nowhere halfway since the Big Bang and grow to become about seventy percent of all the stuff in the universe just to explain the dimness of high redshift galaxies.

The resulting lash-up of a theory has since the early years of the twenty-first century dignified itself with the official title of the Lambda-CDM Standard Model of Cosmology where Lambda refers to dark energy and CDM means cold dark matter. The expanding universe principle has become something of a religion to academic cosmologists, with reputations, livelihoods, and Nobel Prizes at stake. Few dare to take the professional risks involved in criticising it, and contrary publications rarely make it past peer review. As in any religion, passionate belief only becomes invested in doubtful ideas. A Big Bang origin for the universe also brings two further problems, if a big bang did occur it should have created an equal amount of matter and anti-matter, but we cannot find any evidence of the anti-matter. A big bang should not have created a universe of the observed large scale uniformity and homogeneity, to patch this problem cosmologists have introduced the idea that the very early universe underwent an initial period of hyper-extreme expansion called Cosmic Inflation although theorists differ widely and wildly on the possible causes and consequences of this.

Oddly, the most common way to visualise the Big Bang expanding universe begins with the visualisation of a Hypersphere!

Imagine the surface of a sphere, a spherical balloon provides a good example. Place dots all over the surface quite close together, now inflate the balloon and watch every dot move away from every other dot as the balloon expands. The dots represent galaxies, the balloon's surface represents space expanding. The dots will also get a bit bigger, but the galaxies do not because their own gravity holds them together and stops them expanding. To visualise the big bang, imagine that the balloon can start from a size so small that no microscope could render it visible. That may well seem ridiculous, but it arises as a consequence of extrapolating the expansion backwards to a big bang.

Of course, an expanding universe would actually expand in three dimensions, not just in the two dimensions of the surface of a sphere but cosmologists use the analogy of the surface of a sphere to make the points that no galaxy has a central or privileged position in expanding space, and that space does not have an edge as it curves round to make a closed surface. Conventional cosmologists use a closed surface to represent a closed three dimensional space because the observable universe appears to consist of a closed or a very nearly closed three dimensional space albeit a supposedly expanding one, such a space qualifies as a Hypersphere.

Mathematically speaking, 'spheres' come in many dimensions, a one-sphere consists of a one dimensional line curved and closed into a simple circle, a two-sphere consists of a two dimensional surface curved and closed to form the surface of a simple ball or globe, a three-sphere consists of a three dimensional space curved and closed to form something we cannot easily visualise and it has the technical name of a Glome. The Glome and all higher dimensional spheres can bear the name 'hypersphere' but in this exegesis hypersphere will refer exclusively to the Glome or three-sphere.

Note that all spheres involve an 'extra' dimension. The one-dimensional line making up a circle bends through a second dimension, the two-dimensional surface making up a globe bends through a third dimension, the three-dimensional space of a hypersphere bends through a fourth dimension. In this sense the hypersphere represents a 'four-dimensional object' although the 'fourth dimension' here will appear as a spacetime curvature rather than as an extra 'direction' to observers within it.

The surface of a ball or globe thus provides a readily visualisable lower dimensional analogy of a hypersphere. Any point on the surface of a sphere has a corresponding antipode point on the other side of the sphere. The north pole has an antipode point at the south pole on the surface of the earth, the antipode of London lies in the sea off New Zealand. The distance to the antipode shows the maximum possible separation of any two points on the surface of a sphere. If the earth had a much greater mass that prevented light from escaping and confined it to traveling round the surface, then in principle, an observer standing on the north pole could see the south pole by looking in any direction, the south pole would appear faintly smeared all around the horizon of the observer.

Any attempt to represent a shape in a lower dimension involves some sort of distortion. The Mercator style projection of the surface of the earth onto a flat map distorts the polar and near polar regions making Canada, Greenland, and Antarctica look exceptionally vast. Cartographers occasionally use polar projection maps which preserve the size of the polar regions but distort the equatorial regions. This kind of map consists of two circular discs, one a 'photograph' of the earth taken from above the north pole and the other taken from above the south pole. The rims of both discs show the same equator so we can place the two discs flat on a surface touching at some point where the geography matches. Now we can roll one disc round the other and find that the geography matches all the way, so the map remains useful when planning journeys across the equator. This rather unusual two-disc type of map translates into a powerful method of visualising a hypersphere in three-dimensions

Imagine two balls placed in contact. Here we concern ourselves with the three-dimensional space inside the balls rather than with their two-dimensional surfaces. Imagine that the two balls can roll around each other's surfaces in any direction. Imagine that an observer resides in the centre of one of the balls. The centre of the other ball represents the observer's antipode because the observer could travel or see in any direction out through the surface of the surrounding ball and back in through the surface of the other ball to its centre as the two balls 'really' lie with their entire surfaces touching, despite that we can only readily imagine them touching at one point. In this visualisation it becomes apparent that the observer could in principle see the antipode point by looking in any direction in three-dimensional space, so it would appear as a faint sphere surrounding the observer. This does not mean the observer occupies a special position. We could make a two-disc representation of the earth's surface by taking 'photographs' from above any two antipode points. I could make one centred on my house and another point in the south pacific, in a hypersphere any observer can centre the universe on herself.

Note that all types of sphere have, in some sense, 'more room inside'. Flatlanders living on a spherical surface would find that circles on that surface have more surface area than expected and that the radius of a circle seems longer than expected because it has to go over the curve of the surface. A gigantic circle of land on the surface of the earth will have a greater surface area than that within a similarly gigantic circle on a properly flat plane. The maximum amount of 'radius excess' occurs when the circle goes right round a great circle on the spherical surface, for example its equator. To Flatlanders the apparent radius would then equal half the circumference. An analogous effect occurs with a hypersphere, it contains more three-dimensional space than expected by any observer who looked at it from the outside and thought it consisted of a sphere. Within a hypersphere all 'straight' paths from a point to it antipode have the same length, so from the outside the antipode length equals half the circumference.

In Hypersphere Cosmology the observable universe constitutes the entire universe. Observable here means everything as far as the antipode, of course we cannot see everything in detail out to the antipode distance, but we can see some of the bigger galaxies at near antipode distance.

In Hypersphere Cosmology the universe has exactly enough mass/energy to provide the spacetime curvature to close it gravitationally into a hypersphere at its vast size, thus it consists of a black hole. If it had less mass, it would have a smaller size, if it had more mass it would have a greater size. It has exactly enough mass for it size not by coincidence but because of the basic hyperspherical geometry of cosmic scale spacetime. **(1)**

We usually think of black holes as regions of space containing a great deal of highly compressed matter at enormous densities, however really vast black holes can have low densities inside because black holes arise not when the mass to volume ratio reaches a certain level, but when the mass to radius ratio reaches a certain level. As the volume and mass go up by the radius cubed it only takes a few atoms per cubic metre to close a universe the size of the observable universe. A hypersphere the size of the observable universe can consist mostly of space. Astronomers have noticed that the observable universe does seem to contain roughly enough ordinary matter to close it at its observable size, but Big Bang cosmologists dismiss this as 'a mere coincidence' and opine that we inhabit a universe which has by now accelerated far past the size we can observe.

Gravity or spacetime curvature always manifests as an acceleration. Throw a ball upwards and the earth's gravity decelerates it until it momentarily stops going upwards and then it accelerates back towards the ground.

In Hypersphere Cosmology the universe has a small positive spacetime curvature which we can detect as an exceedingly small deceleration in any direction.**(2)** This deceleration explains why space probes sent on voyages around our solar system gradually slow down over long distances, as observed in the Pioneer Anomaly.**(8)** The effect remains small and hard to measure even at solar system distances but at cosmic distances the deceleration influences the light flying between galaxies. The deceleration cannot slow the light down, but it does remove energy from it causing a redshift.**(6)** Thus, the redshift of light from distant galaxies does not arise from recession velocity, it arises purely in response to the distance the light has travelled against the spacetime curvature. Such a 'gravitational redshift' should not occasion theoretical alarm, with a spectroscope we can easily observe a redshift in the light that has climbed out of a heavy star. The Redshift Equation of Hypersphere Cosmology shows the relationship between redshift and the distance to antipode, it does not reveal either distance or antipode distance directly. For these we also need to look at what the positive spacetime curvature of a hyperspherical universe does to our perceptions of distances.

Light follows the spacetime curvature. Proof of General Relativity came with the observation that the light from distant stars curves as it passes close by the sun. (Observers had to wait for a solar eclipse to darken the sun so that they could see this effect.)

The hyperspherical universe acts as giant lens, bending the paths of light rays which travel across it. To understand this effect, consider again either the spherical surface or the two-ball representation of a hypersphere, in both cases the antipode point of any observer becomes spread out right around the horizon, (a circular horizon in the spherical surface example and a spherical horizon in the two-ball example). In both cases the antipode will appear very dim because it has become so spread out. In a hypersphere, very distant luminous objects will appear both larger and dimmer, the increase in apparent size will pass barely noticed because such objects rarely appear as more than point sources even when magnified substantially, but the dimness will register as a considerably reduced apparent magnitude of luminosity.

The Perlmutter et al investigation into the relationship between redshift and distance led to the invention of dark energy because it assumed that spacetime has a Euclidean Geometry in which apparent luminosity falls off with the inverse square of distance. So, if you look at two objects of the same intrinsic brightness and one looks only a quarter as bright then it must lay twice as far away. Space with a Hyperspherical Geometry will make distant objects look much dimmer than they would at the same distance in space with a Euclidean Geometry. So, if observers make the mistake of assuming the universe has a Euclidean Geometry, they will conclude that such objects lay much further away than they do.

Redshift has a logarithmic scale. In Big Bang cosmology a redshift of 1 characterises light coming from events occurring at about half the time since the big bang, a redshift of 10 characterises light coming from events fairly soon after the big bang whilst any light from the big bang itself would have a theoretically infinite redshift making it unobservable. In Hypersphere Cosmology a redshift of 1 characterises light from an event with distance of halfway to antipode whilst higher redshifts characterise light from closer to it, and light coming from the antipode itself would have an infinite redshift.

Now the Perlmutter investigation looked at supernovae exploding in distant galaxies out to redshifts around 0.8, the current limit of observation of such events. These correspond to almost halfway back in time to the big bang or to almost half the antipode distance in hypersphere cosmology. Such supernovae have a brightness which suggests that they must apparently lie about twenty billion light years away if space has a Euclidean geometry. By extrapolating the relationship between such apparent distances and redshift, the big bang itself would seem to now lie about 13.8 billion years away in time and perhaps hundreds of billions of light years distant in space, if not at an infinite distance. Thus, in Big Bang Cosmology the universe expands at far more than lightspeed but according to theorists this does not really matter because it involves space expanding rather than objects travelling beyond lightspeed.

Alternatively, the application of the Hypersphere Lensing Equation **(16)** to the apparent distances of the supernovae yields a set of actual distances for the supernovae, and those at a redshift of around 0.8 lie about six billion light years away. The subsequent application of the Redshift-Distance Equation to the actual distances always recovers an antipode distance close to thirteen billion light years on average, (brightness measurements have moderate uncertainties).

Thus, Hypersphere Cosmology and Big Bang Cosmology paint radically different pictures of the universe by interpreting the same observational data in quite different ways.

To decide between them we need to consider how each theory deals with some of the other phenomena that the universe exhibits.

In Hypersphere Cosmology the universe has spin as Gödel thought, and this spin supplies exactly enough centrifugal acceleration to balance the centripetal acceleration of the universe's positive spacetime curvature.**(5)** It must balance because if the universe started to contract, its spin would increase, and the size would go back up. In effect it does what Einstein intended with his cosmological constant. Neither Gödel nor Einstein had fully investigated the properties of the hypersphere that they thought the universe consisted of and they seemed to think that if it had a spin it would spin like an ordinary ball.

A hypersphere of billions of galaxies spread thinly like a dust cloud can have a 'spin' in which galaxies all spin around each other about randomly orientated planes in a manner that we can call a Vorticitation. Imagine a ball of dust in which all the dust motes move at speed but cannot escape from their collective gravity. Every mote of dust will orbit on a circular path that will pass inside of the orbital path of every other mote. If the dust motes lay thinly spread and move slowly in relation to their size, collisions will remain exceedingly rare. Here the dust motes represent the galaxies and despite their vast size they lie at great distances from each other and despite their substantial velocities they will not appear to move much to human observers to whom the angular separations of galaxies will only change by hundredths of an arcsecond per century. Such randomly orientated circular paths within a hyperspherical space go by the name of Hopf Fibrations. Animations of Hopf Fibrations (albeit in reduced dimensions) exhibit an exquisite beauty. The vorticitation of the universe does not confer any overall angular momentum upon it because the galaxies rotate around randomly orientated planes, but it does confer a basic rotational component to spacetime because everything in it has both a velocity and an acceleration towards the whole.

As Gödel noted in his 1949 paper: 'Matter everywhere rotates relative to the compass of inertia with an angular velocity of twice the square root of pi times the gravitational constant times the density'.

This rotation boosts the rotation of anything rotating in any direction within the universe. At atomic or planetary scales this has an immeasurably small effect, but it has a noticeable effect on galactic rotation, and it will have an effect on black holes.

Conventional theory predicts that a disc galaxy should rotate in a Keplerian fashion with the outer stars of the disc taking much longer to complete their orbits around the centre than stars nearer the centre, much as in the same way that whilst the Earth orbits the Sun once per year, distant Pluto takes 248 years. Galaxies do not rotate like that, beyond the bulge of the central disc the stars at all distances tend to have equal orbital periods, galaxies rotate more like solid discs. Now if we calculate how fast each part of a galaxy should rotate and then simply add on an extra angular velocity from Gödel's equation using a density calculated by including the entire volume of the surrounding spherical gas halo,**(9)** then we get a prediction of the rotation which matches the observed rotation without having to add dark matter as a fudge factor to balance the books.

As closed areas of spacetime, all Black Holes consist of Hyperspheres. All of the material within a hypersphere or black hole has a randomly orientated orbital velocity of lightspeed although the black hole has no overall angular momentum, despite that any accretion disc that forms around it will tend to lie in a particular plane – that of the surrounding galaxy for a large hole. Consequently, the material within a black hole cannot contract into a smaller space without its orbital velocity exceeding lightspeed, so black holes cannot collapse internally into spacetime singularities of infinite density but no size.

The extra angular velocity that the entire universe adds to anything rotating within it will tend to push the outer regions of a black hole towards a rotation speed in excess of lightspeed.**(10)** As the structure of spacetime does not permit superluminal velocities, mass and energy will become shed outwards from the event horizon in the same way as predicted for Hawking Radiation. Whilst Hawking Radiation emission increases with inverse proportion to the size of a black hole, this form of its emission increases in proportion to the size of the hole.

These two effects of the rotation of the universe on the behaviour of black holes have profound implications for the history and future of the universe. Neither increasingly large black holes nor spacetime singularities will eventually suck in all the matter in the universe. The material that falls into black holes eventually becomes recycled back into space.

The Big Bang theory predicts that the universe began with an eruption of energy rather than of matter and that this energy fairly rapidly condensed into matter, but in doing so it should have created an equal amount of anti-matter at the same time. We can observe occasional anti-matter particles (the potassium in a banana emits about one anti-electron per hour), and we can make particle-antiparticle pairs out of energy in particle accelerators, but we cannot observe the mighty blasts of radiation from matter-antimatter annihilations that should occasionally occur if the universe contains large clumps of both.

Hypersphere cosmology predicts that the matter at the antipode of any observer will consist of anti-matter with respect to the observer. The Occultaris part 2 will eventually present a description of matter entirely in terms of spacetime geometry to clarify this point, but for now consider the consequences of moving anything to the antipode.

Observers who travel to the other side of the world become rotated by 180^0 and arrive there upside down with respect to their starting positions. Observers who travelled to the antipode of a hypersphere would arrive rotated by 180^0 in all three spatial dimensions, and hence in complete mirror image form. As the universe consists of a spacetime hypersphere, temporal orientation also becomes rotated by 180^0. Now the spacetime geometry model of particles predicts that an antiparticle simply consists of a particle completely turned around in all directions of space and time, so observers traveling to their antipodes would arrive in a form of matter anti to the form in which they set off. This keeps all types of matter of opposite type safely apart on the 'other' side of the universe, and as matter and antimatter both emit and absorb light and radiation in exactly the same way, the antimatter remains hidden in plain sight. The vorticitation of the universe means that matter turns into anti-matter and back again as it rotates around the universe once every 26 billion years.

Soon after its discovery, the Cosmic Microwave Background Radiation (CMBR) became interpreted as the cooled remains of the intensely hot fireball of the early universe in the Big Bang model.

In Hypersphere Cosmology the CMBR originates not from a time close to the Big Bang but from a place near to the observer's antipode. The CMBR may consist of redshifted near-antipodal starlight that has reached thermodynamic equilibrium with the thin intergalactic medium, but Hyperspherical Lensing raises another possibility:

Widely spatially separated observers within the universe may see a quite different CMBR or perhaps none at all, because the CMBR originates from the antipode positions they once occupied.

That hypothetical galaxy at our antipode may actually consist of our own and the CMBR may represent light emitted from it nearly thirteen billion light years ago, now focussed back on us by the lens of the great hypersphere of the universe, and highly redshifted on its epic journey.

In other words, this galaxy has taken a thirteen billion light year path to its present location and all the unabsorbed starlight from it that left it thirteen billion light years ago now reconverges upon it in much redshifted form. See the diagram near to the end of this appendix.

Astronomers estimate the age of our galaxy as very close to the age of the universe itself in Big Bang terms. That it could have formed so quickly proves something of an embarrassment to Big Bang theorists who hold that the universe has evolved towards its presently observed structure from something that once looked far different.

The evidence for one-way cosmological evolution remains mixed. The entropy of a vast Glome Hypersphere may remain constant as a function of its hypersurface area. **(11)** On the exceptionally large scale the universe needs only the ability to break neutrons to maintain constant entropy. Neutrons that escape the confines of atomic nuclei or neutron stars disintegrate spontaneously. Very distant parts of the universe appear to contain structures far too large to have evolved in the Big Bang timescale.

Physicists have long remained intrigued by Mach's Principle. Broadly and vaguely stated it says that the inertia of any object should depend on the large-scale structure of the rest of the universe. Many physicists believe it should apply to the universe although it does not seem likely to in a universe which began with an enormous if not infinite density that has now dropped to a low level.

Mach's Principle can only work in a universe of constant size and density. Strong evidence exists to show that the gravitational constant and inertial masses have remained constant for billions of years. Hypersphere Cosmology can give Mach's Principle a precise mathematical formulation. **(7)**

The Hypersphere Cosmology model seeks to replace the standard Lambda Cold Dark Matter (LCDM)/Big Bang cosmological model with something approaching its exact opposite. In Hypersphere Cosmology the universe does not expand, it remains as a finite but unbounded structure in both space and time with spatial and temporal horizons of about 13bn light years and 13bn years. The Hypersphere Cosmology universe does not collapse because its major gravitationally bound structures all rotate back and forth to their antipode positions over a 26 billion year period about randomly aligned axes, giving the universe no overall angular momentum and no observable axis of rotation.

The small positive spacetime curvature of the Glome type Hypersphere of the universe has many effects, it redshifts light traveling across it, it lenses distant objects making them look further away, it prevents smaller hyperspheres persisting indefinitely or singularities forming within black holes, and it gradually causes black holes to eject mass and energy.

The universe will appear to observers as having the flux from distant sources in stereographic projection due to the geometry of the small positive spacetime curvature.

In terms of the enormously dimmed flux from very distant sources the antipode will appear to lie an infinite distance away, and beyond direct observation, even though the antipode of any point in the universe lies about 13bn light years distant.

The antipode thus in a sense plays the anti-role of the Big Bang Singularity in Lambda Cold Dark Matter (LCDM) cosmology. We can never directly observe either, but instead of an apparently infinitely dense and infinitely hot singularity a finite distance away in the universe undergoing an accelerating expansion in space and time, Hypersphere Cosmology posits an Antipode that will appear infinitely distant in space and time and infinitely diffuse and cold, even though actual conditions at the antipode of any point will appear broadly similar on the large scale for any observer anywhere in space and time within the hypersphere.

Both Hypersphere Cosmology and LCDM-Big Bang can both model many of the important cosmological observations but in radically different ways. HC has more economical concepts, as a small positive spacetime curvature alone can account for redshift without expansion, the dimming of distant sources of light without an accelerating expansion driven by dark energy, and it also offers a singularity free universe.

Neither model really explains where the universe 'came from', but we have no reason to regard non-existence as somehow more fundamental than existence. Privative concepts do not explain the observed conditions of existence.

Hypersphere Cosmology has the bonus that it explains the 'Large

Number Coincidences' noticed by Eddington and Dirac. The huge number 10^{40} seems to occur frequently, particularly in relation to the ratios between cosmic and particle phenomena. Big Bang theorists usually dismiss this as a mere coincidence apparent only at this epoch of the universe's evolution. In Hypersphere Cosmology the large numbers, typically 10^{20}, 10^{40}, 10^{60}, 10^{80}, and sometimes even 10^{120}, all arise naturally and persist for all time because of the ratio of the universe to the Planck scale. **(11-15)**.

Key to algebraic symbols

M = Mass of universe. m = Mass. G = Gravitational constant. c = Lightspeed. L = Antipode distance of universe. d = Distance. p = Density. E = Energy observed. ë = Wavelength. f = Frequency. ù = Angular velocity. X_0 = Observed value of X. X_e = Expected value of X. A = Acceleration due to spacetime curvature of universe. a = Acceleration. v = Velocity. F = force. r = Radius.

Key to mathematical notation.

e3 or 10^3 = 10 x 10 x 10 = 1,000. 10^6 = 1,000,000. Etc. The index gives the number of zeros.

e-2 or 10^{-2} = 1/100. e-3 or 10^{-3} = 1/1,000. Etc. Negative indices show hundredths, thousandths, etc.

All data presented in SI units, metres, kilograms, seconds.

Formulae:

Schwarzschild Black Hole. $\frac{2Gm}{r} = c^2$ This derives from General Relativity and shows the ratio of mass to radius required to create an escape velocity of lightspeed, and hence a Black Hole.

Gödel Rotation. $\omega^2 = 2\sqrt{\pi Gp}$ This derives from General Relativity and shows the angular velocity arising from any given density of matter.

Hypersphere volume $= 2\pi^2 r^3$ or $\frac{2L^3}{\pi}$ where r = externally measured radius. $\pi r = L$

Velocity, distance, and acceleration. $v^2 = 2da$ This shows the velocity arising from the application of an acceleration over a distance.

Angular velocity and frequency. $\omega = 2\pi f$ This converts angular velocity (in radians) into frequency of rotation.

Redshift = Z. $Z = \frac{\lambda_o}{\lambda_e} - 1$ The redshift shows the ratio of the observed wavelength to the expected wavelength with one subtracted from it so that the scale starts at zero rather than one.

The Equations of Hypersphere Cosmology.

(1) $\frac{2GM}{L} = c^2$ Black Hole as a Hypersphere. As a closed area of spacetime, any Black Hole also consists of a Hypersphere.

(2) $\frac{2GM}{L^2} = A$ A Hypersphere will have a characteristic positive spacetime curvature which appears as a small deceleration of all linear motion within it.

(3) $\frac{GM}{L^2} = \frac{A}{2} = \frac{c^2}{2L}$

(4) $A = \dfrac{c^2}{L}$ Centripetal Acceleration.

The Universe thus has the following size and curvature.

Antipode length, L = 1.23 x 10^{26} metres. 13 billion light years. See **(17)**

Total mass, M = 8 x 10^{52} kg. Eighty Octillion Tonnes.

Curvature as acceleration, A = 7.317 x 10^{-10} metres/second^2.

(5) $\omega = 2\sqrt{\pi G p}$ $\omega^2 = \dfrac{4\pi^2 GM}{2L^3}$ $\omega^2 = \dfrac{\pi^2 c^2}{L^2}$

$\omega = \dfrac{\pi c}{L}$ $2\pi f = \dfrac{\pi c}{L}$ $f = \dfrac{c}{2L}$

$A = \dfrac{c^2}{L}$ The 'rotation' of the universe gives rise to a centrifugal acceleration which naturally balances its centripetal acceleration.

(6) $\dfrac{1}{Z+1} = \dfrac{\lambda_e}{\lambda_o} = 1 - \dfrac{dc^2}{2GM} = 1 - \dfrac{dA}{c^2} = 1 - \dfrac{d}{L}$

The wavelength of light remains inversely proportional to its energy. Deceleration over distance removes energy from light creating a Redshift. From this we can derive the following three useful equations.

$z = \dfrac{1}{1 - \dfrac{d}{L}} - 1$ $d = L\left(1 - \dfrac{1}{Z+1}\right)$ $L = \dfrac{d}{1 - \dfrac{1}{Z+1}}$

(7) $\dfrac{2M}{L} = \dfrac{c^2}{G}$ $\dfrac{2GM\,m_g}{Lc^2} = m_i$ Mach's Principle.

The ratio of an objects gravitational mass m_g to its inertial mass m_i arises from the mass and size of the entire universe and these determine lightspeed and the gravitational constant.

(8) $v_2^2 = v_1^2 - dA$ Pioneer Anomaly.

The spacetime curvature A slows the velocity of objects moving long distances.

(9) $v(r) = v_i(r) + 2\sqrt{\pi G \rho}\, r$

The rotational velocity of a galaxy at radius r equals the inertial velocity expected from the ordinary mass distribution plus the Gödel angular velocity calculated using the entirety of the mass within the spherical gas halo.

(10) $F = ma - mA$ $ma = \dfrac{Gmm}{r^2} + mA$ $a = \dfrac{Gm}{r^2} + A$ $v = \sqrt{\dfrac{Gm}{r} + rA}$

The spacetime curvature A modifies the orbital velocity of large black holes causing them the slough off mass and energy.

(11) $\dfrac{2M}{m_p} = \dfrac{L}{l_p} = \dfrac{T}{t_p} = \dfrac{a_p}{A} = U\,10^{60}$

The ratios of Planck units to the size of the Universe's dimensions. U = what we can call 'The Ubiquity Constant'.

(12) $\quad S \vee I \dfrac{L^2}{l_p^2} 10^{120} \qquad \dfrac{L^3}{l_p^3} 10^{180}$

If the Bekenstein-Hawking Conjecture that the Entropy and Information content of a black hole corresponds to its surface area in Planck units holds for the entire universe then the universe has only one unit of information per 10^{60} Planck volumes or one unit per $\sqrt[3]{U} = 10^{20}$ Planck lengths.

(13) $\quad \Delta E \Delta t \hbar \sqrt[3]{U}$

The effective level of uncertainty/indeterminacy in the universe begins at twenty orders of magnitude above the Planck scale.

(14) $\quad l_p \sqrt[3]{U} \quad t_p \sqrt[3]{U}$

We do not actually observe any phenomena smaller than twenty orders of magnitude above the Planck length or the Planck time.

(15) $\quad m_n = \dfrac{m_p}{\sqrt[3]{U}} \quad N_n = U^{\frac{4}{3}}$

The largest stable matter particles, (neutrons or the proton-electron considered as a pair), have a mass twenty orders of magnitude below the Planck mass. Consequently, the universe contains 10^{80} baryons.

(16) $\quad d = d_o 1 - \sqrt{1 - \dfrac{1}{Z+1}}\quad$ The Hyperspherical Lensing Equation. Derived from the following geometric considerations.

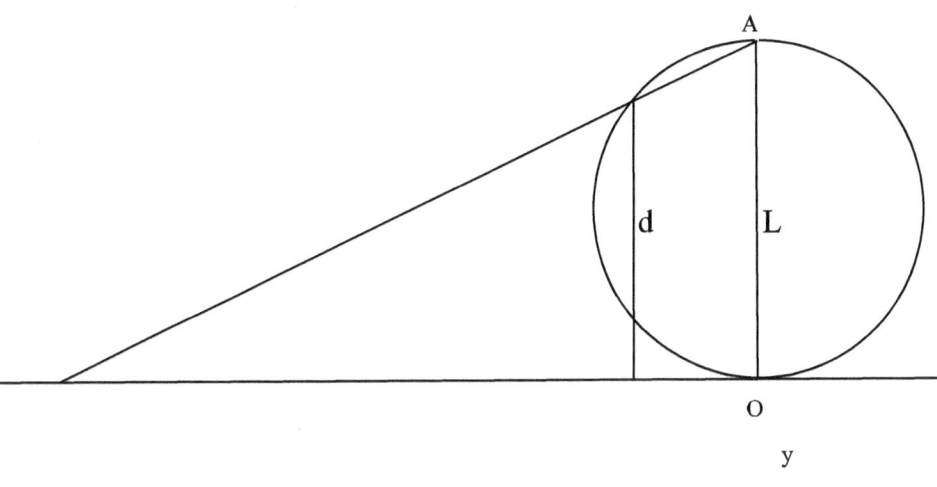

O represents the position of the observer, A the observers antipode. Note that in a hypersphere, represented by the circle, the distance L equals half the circumference. Then by similar triangles: -

$$\dfrac{x-y}{x} = \dfrac{Jy_0}{Jy_o - Jy} = \left(\dfrac{d_o - d}{d_o}\right)^2 = \dfrac{d}{L}$$

Jy = Flux in Janskys. Jy_o = Observed flux calculated from apparent magnitude.

$$d = d_o 1 - \sqrt{\dfrac{d}{L}}$$

$$d = d_o 1 - \sqrt{1 - \dfrac{1}{Z+1}}$$

The universe thus appears in Stereographic Projection.

(17) $$L = \frac{d_o \left(1 - \sqrt{1 - \frac{1}{Z+1}}\right)}{1 - \frac{1}{Z+1}}$$

By calculating Apparent distances for type 1A Supernovae based on an inverse square diminution of flux with distance and then using their redshifts in hyperspherical lensing to obtain their actual distances and then using their redshifts in the redshift-distances equation to calculate the antipode length of the universe, a value of L close to 1.23×10^{26} metres appears in all cases.

The universe thus consists of a vorticitating hypersphere of constant size, finite and unbounded in both space and time.

The CMBR

The Cosmic Microwave Background Radiation has become interpreted as evidence for the Big-Bang expanding universe LCDM standard cosmological model.

The CMBR consists of electromagnetic microwave radiation that peaks at a frequency corresponding to blackbody radiation with a temperature of 2.7^0 Kelvin above absolute zero. It comes to observers here from all directions in space.

Standard cosmology interprets the CMBR as relic radiation leftover from a primeval fireball about a third of a billion years into the supposed expansion of the universe from a spacetime singularity or near singularity. At this time the universe had supposedly expanded to a radius of a third of a billion light years and cooled to a temperature of about 3,000K, at which point it deionised from a proton-electron plasma and allowed photons to pass freely through it. Such photons which allegedly had very high energies and short wavelengths, subsequently became much lower energy longer wavelength photons due to the expansion of space and they now appear to us as the microwave background radiation.

We cannot see the CMBR with the naked eye, but if we could, the night sky would not appear dark between the stars because about 400 CMBR photons per square centimetre per second impinge from all directions, they represent a significant cosmic phenomena that demands an explanation.

In Hypersphere Cosmology, all the electromagnetic radiation that does not become absorbed on its journey will eventually return to its point of origin after about 13 billion years but in a much redshifted form due to gravitational redshifting and lensing both caused by the small positive hyperspherical spacetime curvature of the entire universe.

Cosmologists estimate that our galaxy, the Milky Way, has an age alarmingly close to the supposed age of the universe itself, somewhere around the thirteen billion year mark. They remain adamant that it cannot possibly have an age greater than their estimated age of the universe. Yet the Milky Way contains the so-called Methuselah star (HD 140283), whose apparent longevity does seem to severely challenge standard the cosmic age limit.

The average surface temperatures of all the billions of stars in a galaxy or a galactic cluster adds up to something resembling a blackbody radiation source at about 3,000K when seen from a cosmic distance.

Thus, the CMBR we observe in this region of the hyperspherical cosmos may well have originated from this region and reconverged back here in highly redshifted form thirteen billion years later. Observers in a deep intergalactic void may not observe any CMBR at all.

The following two diagrams show the combined effects of Hyperspherical Lensing and Hyperspherical Vorticitation. Hyperspherical Lensing brings electromagnetic radiation to a focus at the antipode point to its emission. Hyperspherical Vorticitation moves the source of emission to its antipode point in the same period. The first diagram shows this effect in the reduced dimension 'surface of a sphere' representation of a hypersphere, and the second shows it in the reduced dimension 'two-ball' representation of a hypersphere. Thick arrows represent the movement of the source due to vorticitation; thin arrows represent the paths of emitted radiation.

    Sphere model   Two-Ball model.

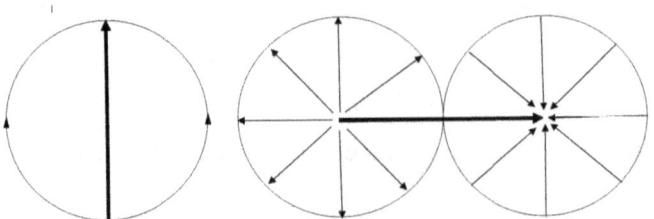

Hyperspherical vorticitation will send this galaxy to its antipode point in the universe in a period of 13 billion years. Light emitted from this galaxy will also travel to the same antipode point in the same period and appear as incoming light. The galaxy travels in just one direction, the light travels in every direction. The incoming light will appear highly redshifted from stellar surface temperatures to 2.7^0 K.

Hyperspherical lensing will make the incoming light appear to have come from a surrounding sphere with a diameter very much larger than the antipode distance but compressed into a smaller field of view and thus multiply imaged. This will effectively smear out temperature variations and absorption and emission lines to yield a CMBR with blackbody radiation characteristics.

Quasar Distribution

Quasars seem to consist of active galactic nuclei where the swirling in-fall of matter towards the central black hole of a galaxy causes the emission of massive amounts of radiation.

We have observed about a million quasar galaxies amongst the estimated 100 or more billion galaxies. As quasars seem to emit such unfeasibly vast amounts of energy, theorists have concluded that they must emit it mainly as beams which project out of the spin axis of the black hole and its accretion disc, rather than spherically in all directions. In this they behave very much like Pulsars which consist of rotating neutron stars with radiation beams coming out of their magnetic poles.

Thus, we can only detect Quasars that have one of their radiation beams pointing roughly towards us.

The closest detected Quasar to us lies about 0.6 billion light years away and the apparent density of Quasars in the universe seems to increase markedly with distance. This has become interpreted as supporting evidence for a big-bang expanding universe theory in which earlier galaxies had more material close to their black holes and behaved as Quasars until the black hole had consumed it. This assumes that galaxies have tended to undergo a one way evolution from having an active nucleus to having an inactive one.

Hypersphere cosmology asserts that galaxies recycle and reform themselves endlessly and that galaxies at all stages of change will appear over all observable space and time, and that observers will find more quasars at long distances because of the positive curvature of spacetime.

If we assume that the spin axes of widely separated galaxies lie randomly orientated with respect to each other, then we should only expect to observe a small proportion of all quasars. However, positive spacetime curvature has a lensing effect on light and other forms of radiation that allows us to see more of the surface of an object than in flat spacetime, and the effect increases with distance.

For example, observers can see 70% of the surface of a neutron star rather than just 50% of it because of the curvature of spacetime around such a dense object.

In a hyperspherical universe we could in principle see 100% of the surface of an object at antipode distance. It would appear as spread around the observer's entire spherical horizon. If such an object consisted of a Quasar, the observer could in principle see both polar beams from it coming apparently from opposite sides of the universe.

Thus, the further away a Quasar lies from an observer, the higher the probability that spacetime curvature will 'tilt' the rotation axis of that Quasar in the observer's direction.

The Starlit Sky

Away from the glare of our burning cities
The sky at night looks black as pitch betwixt the stars
Olbers said that cannot be
In a universe of infinity
Yet look again with a microwave eye
And it's all aglow betwixt the stars
Is this the relic of a mighty blast
From distant aeons in the past
Or does the light from here
Chase us round a cosmic hypersphere
Coming back a little tired and cold
After all, it's very old.

Index of Questions

Interview 1
1) Cosmology, The Magical origins of Hypersphere Cosmology.
2) Magic and Science.
3) Magical techniques for Science.
4) Magical Models, Orgone Energy, Quantum Transactional.
5) Science versus Magic.
6) Science versus Magic.
7) Religion in Science.
8) Magic, Quantum, Cosmology.
9) Pointy hats, Newton.
10) Academia, Information technology.
11) Reversible time.
12) Hypersphere and Existence.
13) Cosmological and Quantum experiments.
14) Maths.
15) Quantum interpretations
16) Space travel, genetic modification.
17) Ceremonial Magic.
18) Chaos magic
19) Magic and Mind.
20) Styles of magic.
21) Magic 1985-95.
22) Long term magic effects
23) R A Wilson, magical developments.
24) Maybelogic, Arcanorium.
25) R A Wilson – impressions.

26) Biography- University.

27) Study in the Himalayas.

28) Magic and Parenting.

29) Wisdom or Power?

30) Magic for schoolchildren?

31) Chaos Magic in Business.

32) Leary and Wilson's ideas on consciousness.

33) Psychedelics?

34) Aliens?

35) Futurology?

36) Astrology?

37) Book recommendations?

Interview 2

38) Yoruba Practices.

40) Voodoo, Theology & Politics.

41) 'Superhuman' feats.

42) Botanic hallucinogens.

43) Free-Will and computers.

44) Theories of Mind.

45) Panpsychism.

46) Chapel Perilous Multi-Mind.

47) MacGregor Mathers and Chaos magic.

48) LSD and self.

49) Artificial Intelligence and Artificial Stupidity.

50) Doctor Who.

51) Online Hypersphere debate.

52) JFK and Doctor Who.

53) Freemasonry.

54) The IOT, Wilson, Leary, and Burroughs.

55) Conspiracy Theory.

56) Conspiracy and change.

57) Magical Thinking and Persuasion.

58) Resisting the Dark Arts.

59) Occult Military Intelligence.

60) Knights of Chaos.

61) Parapsychology and Barbed Wire.

62) Danger, Fear, and Boldness.

63) Puberty Rites.

64) Rites of Passage.

65) Feng Shui.

66) Ritual Architecture.

67) Sex Magic.

68) Love Magic.

69) Natural Phenomena, Gods, and People.

70) Self-consciousness.

71) Neuromancy.

72) Gnosis and Neuroscience.

73) Kabala.

74) Judaism & Kabala.

75) Newton, scientist, and sorcerer.

76) Magical Apps.

77) Bad Magic.

78) Horrible Gnosis.

80) Evil.

81) Nazi occultism.

82) The Ice Magic war.

83) Helmut 1.
84) Helmut 2.
85) Helmut 3.
86) Fighting dirty.
87) Combat soundtrack.
88) Helmut 4,
89) Cults.
90) Blavatsky and Race.
91) Christianity and magic.
92) Musical tastes?
93) Situationism? Psychogeography?
94) Gnosis and Cult Manipulation.
95) BITE 1
96) BITE 2.
97) Martial Arts?
98) Twin Towers?
99) Prince Andrew.
100) Monarchy.
101) Political Philosophy?

Interview 3
102) Twin Towers conspiracy?
103) Power and Capital.
104) Guns, Drugs, and Prostitutes.
105) Scandinavia.
106) Survival of the fittest.
107) Brexit.
108) Monarchy.
109) UK Government.

110) Local Government.

110) Boredom and Alienation.

112) Boredom and Alienation 21

113) Complexity and Conspiracy.

114) Scandinavia 2.

115) Constitutional Monarchy.

116) Qualities of a modern Monarch.

117) Direct Democracy?

118) Communism?

119) Anarchism?

120) **Biography. See below.**

121) Schooling?

122) Impressions of School?

123) Favourite Subjects.

124) Happy Schooldays?

125) Part time jobs.

126) 'Sixties' Counterculture, Hippies.

127) Magic and Drugs?

128) Mythological fascinations.

129) Getting into Magic.

130) Establishing Chaos Magic.

131) Why I did it.

132) The reception of the books.

Interview 4

133) Magical effects.

134) UK Independent publishers.

135) UK occult fanzines.

136) New Equinox Magazine.

137) The Memsahib.
138) Aleister Crowley's influence.
139) Crowley's influence 2.
140) Objections to Crowley.
141) Business career.
142) Arcanorium College.
143) Arcanorium College 2.
144) Magical ideas throughout history.
145) Platonic Pagan Monotheism.
146) Spirits and Deities.
147) Chaos Magic and the era.
148) Chaos magic and the era 2.
149) Paul Huson, Mastering Witchcraft.
150) Exorcism and In-sorcism.
151) Attempted Exorcism.
152) Criminal Magic.
153) Criminal Magic 2.
154) Giordano Bruno and Dr John Dee.
155) Austin Spare's influence.
156) Spare and Crowley.
157) Spare's sources.
158) Spare's book.
159) Teaching career.
160) Teaching career 2.
161) Insights from Teaching.
162) Travel.
163) Squatting in London.
164) Magical characters, Charly Brewster, Gerald Suster, Amado Crowley, Lionel Snell.

Interview 5
165) Warp Drive.
166) Near Space Uses.
167) Moon Base?
168) Mars Colony?
169) Generation Ships?
170) Celtic Underworld.
171) Eureka and Humour.
172) IOT Rituals.
173) Newton, Alchemy, Gravity.
174) Cosmology.
175) Esoteric Medicine?
176) Adventures in Afghanistan and India.
177) Hinduism
178) Imperialism and Esoterics.
179) Bodily Esoterics.
180) Gurus.
181) Hypersphere Cosmology and Quantum Hyperspheres?
182) Azathoth and Yog-Sothoth.
183) Heresy, Iconoclasm, and Anathema.

Interview 6
184) The Magical Diary.
185) My Diaries.
186) My Diaries 2
187) My Diaries 3
188) My Diaries 4
189) My Diaries, Will and Imagination.

190) Dream Diaries.

191) My Dream Diaries.

192) Dreamwork.

193) Dreamwork 2.

194) Sculpture.

195) Sculpture and states of mind.

196) No-Mind.

197) No-Mind 2.

198) No-Mind 3.

199) Visualisation.

200) Visualisation 2.

201) Constitutional Presidency and Monarchy.

202) Royal behaviour.

203) Social organisation, Chaocracy?

204) Proportional representation?

205) Brexit.

206) Chaocracy 2.

207) Brexit, nationalism, racism.

208) Nationalism and Brexit.

209) History Wars.

210) Environment, Growth, and the Future.

211) Three Occult Revivals.

212) Knights of Chaos.

213) Knights of Chaos 2.

214) Knights of Chaos 3.

215) Magical Entities.

216) Entities and Consciousness.

217) God-Forms.

218) Obsession.

219) Entities and Aliens.

220) Magical languages.

221) Magical languages 2.

222) Ouranian Barbaric.

223) Magical Incantations.

224) Ouranian Barbaric 2.

225) Ouranian Barbaric 3.

Interview 7

226) Universal Basic Income.

227) Nuclear Power.

228) Virtual Reality.

229) Ritual Dance.

230) Ritual Dance 2.

231) Faith?

232) Truth?

233) Druidry.

234) Internet uses and perils.

235) Virtual Tele-Dildonics and Snuff Movies.

236) Cattle Mutilations and Aliens.

237) Alien Communication.

238) Space Germs.

239) Is consciousness evolving?

240) Wilson-Leary 8 circuit brain model.

241) White Magic and Evil.

242) Xenobiology.

243) Xenobiology 2.

244) Multitasking and Education.

245) Transcendental?
246) Wilson-Leary 8 circuit brain model 2.
247) Brain-Machine interfaces?
248) Wilson-Leary 8 circuit brain model vs Lovecraftian Perspective.
249) Tactile perception of images.
250) CGI.
251) Brain 'circuits'.
252) War, Crime, and Progress.

Interview 8

253) Meditation and Compassion.
254) GDP and Gross National Happiness.
255) Rotating Steady State Economy. RSSE.
256) Inequalities.
257) Inequality and Productivity.
258) Achieving RSSE.
259) Achieving RSSE 2.
260) Renewable Energy Currency.
261) Renewables technology?
262) Renewables Research.
263) Achieving RSSE 2.
264) Global Grid?
265) Averting eco-catastrophe?
266) Lifestyle changes, pleasure, or virtue?
267) Consensus Reality.
268) Cultural Upheaval and Magic.
269) Coincidences and synchronicities.
270) Psychotherapy?
271) Magical Thinking and Apophenia.

272) Chaos Magic pitfalls.

273) Archetypes.

274) My daily magical practise.

275) My daily magical practise 2.

276) Obsessions.

277) Obsession and Insanity.

278) Intuition.

279) Sex and motivations.

280) Psychiatry.

281) Mass shootings.

282) Demonic Possession.

283) Tantra.

284) Breath Control.

Interview 9

285) National Exorcism & Humour.

286) 'Higher' consciousness.

287) 8 Circuits & Kabala.

288) Humour & Religion.

289) Trickster gods.

290) Magic & Religion.

291) Historical movements in my youth.

292) The Birth of Chaos Magic.

293) The Next Magical Revival?

294) Time Travel?

295) Scrying the Past.

296) A view of America.

297) China, America, and Climate Change.

298) Time Perception.

299) Time Perception 2.

300) Subjective time dilation and contraction.

301) Time & Magic.

302) Monotheism & Magic.

303) English Language and Slangs.

304) Politically Correct Language.

Interview 10

305) Poetry & Slang.

306) Poetry & Magical Thinking.

307) Taboo & Magical Thinking.

308) Dr Who Monsters and Us.

309) Sacred Geometry?

310) Knights of Chaos.

311) Knights of Chaos targets.

312) Maybelogic & Arcanorium Colleges.

313) Seminar Recordings?

314) Communicating with Plants and Animals.

315) Magical equipment and accessories.

316) Time Binding and Information Doubling.

317) Information and Non-Locality.

318) Information and the Universe.

319) Information and the Universe 2.

320) Death.

321) Near Death Experiences.

322) Chaos Magic misconceptions.

323) Wilhelm Reich & Cloud-Busting.

324) Weather Engineering.

325) Scientific R&D.

326) World without Internet.

Interview 11

327) Sexism, Religion, and Civilisation.

328) Time Crystals.

329) Civilisational Collapse.

330) Escape to Space?

331) China and Thorium.

332) Energy and Geopolitics.

333) Energy and Currency.

334) Currency.

335) Currency, Gold, Cryptocurrency.

336) Crowley, sex and drugs and mysticism.

337) Addiction.

338) Sensory Deprivation.

339) Remote Viewing.

340) Governments and the paranormal.

341) Science from Magic?

342) Copenhagen Interpretation?

343) Space telescopes and cosmology.

344) Standard Cosmology Model.

345) Hypersphere Cosmology.

346) Cosmological debate.

347) Physics and Metaphysics.

Interview 12

348) Simulation Theory?

349) Multiple Universes?

350) Extra Dimensions.

351) Three Dimensional Time.

352) Exploring space and time.

353) The functions of humour.

354) Extra Dimensions, Quantum Chess Game.

355) Does Humour Evolve?

356) Has Magic Evolved?

357) What can we know?

358) Language Evolution, Language and Thought.

359) Visual Gnosis.

Interview 13

360) Hypersphere Cosmology and Black Holes.

361) Optimist or Pessimist?

362) Human Nature?

363) Secret of Life?

364) Self-Image?

365) Explain Hypersphere Cosmology?

Index

A

Addiction 232, 233
Aeonics 193
Afghanistan 112, 113, 310
Al-Ghazali 202
Aliens 28, 153, 253
America 44, 147, 181, 196, 197, 314
Anarchism 68, 132
Animism 83, 140, 218
Apophenia 16, 48, 128, 141, 175, 177, 188, 205, 313
Arcanorium College 25, 45, 82, 177, 209, 210, 309
Argentum Astrum 39, 80, 90, 93
Artificial Intelligence 38, 305
Astrology 29, 109, 305
Azathoth 121, 122, 142, 188, 310

B

Baphomet 56, 91, 109, 141, 142, 148, 161, 162, 188
Bergson, Henri 192
Bhagwan Rajneesh 118
Big bang 13, 17, 20, 196, 238, 239, 240, 241, 246, 263, 277, 278, 284
Biology 26, 50, 93, 95, 96, 159, 194
Bitcoin 231
Blavatsky, Helena 54, 57, 307
Bohr, Neils 245
Brewster, Charly 99, 309
Brexit 64, 133, 134, 135, 136, 307, 311
Bruno, Giordano 88, 309
Buddhism 36, 114, 115, 164, 182, 273
Burns, Robert 268
Burroughs, William 40, 41, 306

C

Chaocracy 132, 135, 311
Chaos Magic 7, 13, 16, 32, 40, 51, 77, 78, 83, 86, 115, 122, 124, 174, 175, 192, 193, 209, 219, 237, 254, 258, 305, 308, 309, 313, 314, 315
Chemistry 13, 25, 71
China 119, 139, 197, 198, 228, 229, 314, 316
Christianity 39, 51, 57, 84, 119, 123, 124, 139, 150, 182, 183, 201, 307
Conspiracy 5, 8, 41, 42, 43, 59, 62, 66, 306, 307, 308
Copenhagen Interpretation 236, 237, 316
Crime 162, 313
Crowley, Aleister 8, 9, 29, 39, 44, 74, 77, 79, 80, 81, 88, 90, 91, 93, 100, 102, 125, 126, 127, 162, 178, 192, 231, 232, 233, 309, 316
Crowleyanity 78, 80, 81, 178, 258
Cult 40, 41, 54, 56, 58, 80, 123
Currency 43, 166, 168, 229, 230, 231

D

Dance 147, 148, 174
Davis, Wade 31
Dee, John 88, 309
Democracy 60, 61, 68, 131, 132, 196
Divination 27, 179
Doctor Who 38, 52, 206, 207, 305, 315
Druid 23, 151

E

Economy 72, 165, 168, 169, 197,

224, 228, 229
Einstein, Albert 17, 66, 110, 240, 274, 275, 276, 285
Elder Gods 48, 122, 128, 142, 160, 176, 177
Enchantment 27
EU 43, 64, 133, 134, 135
Evocation 27
Exorcism 86, 87, 127, 185

F
Faith 148, 149, 312
Fascism 43
Fra Choronzon 999 100
Free Will 33
Freemasonry 39, 305
Futurology 194, 226

G
Ganj, Macleod 114
General Relativity 235, 236, 239, 274, 283
Gnosis 48, 50, 52, 91, 107, 148, 180, 183
Gnostic 123, 126
Golden Dawn 36, 39, 40, 74, 79, 80, 92, 139, 191
Grant, Kenneth 74, 90
Gwyn ap Nudd 107

H
Hawking, Stephen 241
Heisenberg, Werner 245
'Helmut' 54, 55, 56, 307
Hinduism 36, 118, 182, 310
Hippy 58, 73, 190, 191
Humor 107, 108, 185, 186, 187, 253, 254, 257
Huson, Paul 73, 74, 86, 309
Hypersphere Cosmology 5, 7, 9, 14, 17, 21, 122, 165, 236, 239, 242, 263, 270, 272, 273, 281, 282, 284, 285, 288, 289, 290, 291, 299, 304, 310, 316, 317

I
Ice magic 54, 55
Illumination 27
India 25, 26, 69, 73, 75, 81, 94, 96, 98, 100, 112, 113, 116, 118, 119, 139, 182, 191, 197, 212, 310
Internet 141, 151
Invocation 27, 175
IOT 13, 40, 41, 54, 56, 76, 108, 177, 306, 310
Iran 112

J
Judaism 51, 118, 201, 306

K
Kabala 50, 51, 84, 85
Knights of Chaos 42, 44, 45, 140, 208, 306, 311, 315
Korzybski, Alfred 213, 214, 261

L
Lamaism 114
Leary, Timothy 27, 40, 41, 153, 155, 158, 305, 306, 312
Levi. Eliphas 74, 139, 141, 161
Liber Null 11, 26, 39, 75, 76
Lovecraft, H.P. 122, 141, 159, 160, 176, 313
LSD 8, 37, 72, 73, 191, 305

M
Magical Thinking 29, 30, 103, 174, 175, 306, 313, 315
Mandrake 211
Mandrake (Publisher) 77
Mathers, Macgregor 29, 36, 39, 40, 79, 84, 191, 192, 193, 305
Monarchy 60, 68, 120, 131
Monotheism 50, 83, 84, 139, 266, 267, 309, 315
Morocco 98, 112
Multiple universes 246, 247

N

Napoleon Bonaparte 148
Nazis 53, 54
Necronomicon 24
Newton, Isaac 16, 18, 19, 51, 66, 107, 109, 110, 191, 276, 306, 310
Norse Runes 143

O

Optimism 264
Other dimensions 249, 255, 256
Ouranian-Barbaric 90, 144, 145
Ouranos 141, 144, 188

P

Pakistan 113
Panpsychism 35, 218, 305
Paranormal 234, 316
Physics 13, 207, 316
Placebo 30
Pope 42, 208, 241
Psychedelic 32, 37
Psychoanalysis 127, 173
Psychonaut 75

Q

Quantum Field Theory 15, 21, 235, 236
Quantum physics 21, 120, 121, 215, 236, 240, 244, 245, 251

R

'Ralph' 55, 56
Regardie, Israel 75, 173
Reich, Wilhelm 219
Renewable 147, 166, 168, 169, 170, 171, 220, 229

S

Schopenhauer, Arthur 140, 192
Schrodinger, Erwin 267
Sculptures 126, 128
Sex Magic 8, 306
Shamanism 25, 35, 140, 218
Sherwin, Ray 74, 77, 81, 88, 89
Snell, Lionel 8, 29, 34, 43, 74, 103, 245, 309
Sorcerers Apprentice 75, 77
Sorcery 30
Spare, Austin Osman 29, 74, 90, 91, 92, 161, 173, 192, 233, 309
Suster, Gerald 8, 75, 80, 101, 309
Synarchy 43, 64, 134
Synchronicities 173, 313

T

Tantra 182, 314
Thelema 80, 81, 100, 101, 102, 104, 124, 178, 258
Theology 31, 305
Thorium 221, 228, 229
Three-dimensional time 251, 252, 263, 264
Tibetan 25, 114, 115, 164
Time travel 194, 252, 253
Transactional Interpretation 16, 19, 121, 237
 of quantum mechanics 215
Transcendental 49, 155, 157, 194, 258

U

USA 23, 41, 106, 196, 197

V

Visualization 111, 130, 147, 177

W

Warp drive 105, 107
Wilson, Colin 24, 25, 27, 35, 40, 41, 42, 74, 82, 155, 158, 209, 214, 304, 305, 306, 312
Wittgenstein, Ludwig 261

Y

Yoruba 5, 31, 36, 305

Z

Zoroastrianism 200

If you enjoyed this book
and want to know more
sign up for free Mandrake monthly book newsletter, here's how:
Visit the
mandrake.uk.net
website
A subscription page should pop-up

or type this link into a browser

http://eepurl.com/THE9P

www.ingramcontent.com/pod-product-compliance
Lightning Source LLC
Chambersburg PA
CBHW071000160426
43193CB00012B/1847